Unfinished Business

Unfinished Business

A BIOLOGIST IN THE LATTER HALF
OF THE 20TH CENTURY

Joseph T. Bagnara

Unfinished Business: A Biologist in the Latter Half of the 20th Century

Copyright © 2013 Joseph T. Bagnara. All rights reserved. No part of this book may be reproduced or retransmitted in any form or by any means without the written permission of the publisher.

Published by Wheatmark®
1760 East River Road, Suite 145
Tucson, Arizona 85718 U.S.A.
www.wheatmark.com

ISBN: 978-1-60494-961-2 (paperback)
ISBN: 978-1-60494-982-7 (ebook)
LCCN: 2013935624

Table of Contents

Preface vii

Introduction xi

1   Some Personal History 3

2   Graduate School 20

3   The Start of a Professional Career 48

4   The Stadium Years 55

5   Going to Europe 65

6   Building a House 80

7   A Sabbatical in Paris 93

8   Some Productive Years 109

9   The Vicarious Enjoyment of Field Biology 121

10   A Second Sabbatical—Italy, Our Second Home 134

11   The Epoch of the Mexican Leaf Frog 157

12   Important Travel in 1977 171

13   The Emergence of Pigment Cell Biology 178

14   Making a Movie 188

15	Italy Revisited	193
16	Again and Again	204
17	Green, Blue, Yellow, and White Eggs	219
18	Teaching	224
19	The Developmental Biology of Pigmentation Patterns	229
20	The Final Years of Research	239
21	Japan and Friends	244
22	The Retirement Years	263
23	Some Passions	280

Epilogue 298

Acknowledgments 301

A Selected Glossary 303

Especially Relevant Research Publications 309

Preface

A book preface is often written after the book is completed, but this one is written at the outset in the way of an apology since it precedes a kind of autobiography. The recounting of one's life, in my eyes, is rather presumptuous. My guilt is assuaged because the more than fifty years of my professional life took place in a unique period that followed World War II when the changes and growth in academia were profound and rapid. I wish to tell about it.

At that time, universities were primarily teaching institutions but, in only a very few years, their research mission expanded such that the creation of knowledge often began to supersede the dissemination of knowledge. Perhaps this was for the better, but I am not so sure. In any case, the recounting of these changes from my viewpoint, in the context of my own creative past, might be found interesting or amusing by current generations.

When I joined the faculty of the University of Arizona well more than fifty years ago, the times dictated that I was to teach and contribute in the realm of public service to the people of Arizona. Any efforts toward original research were incidental in the eyes of the university administration, and certainly, such activities from the College of Liberal Arts were not expected by the state legislature. In more progressive state universities, such as those in California, or my own PhD alma mater, the University of Iowa, original research was respected and revered. Fortunately, at the time I was hired in 1956, some of the younger faculty had begun to press for the addition of young, more research-oriented replacements for vacant faculty positions.

When the head of the Department of Zoology retired at the end of the 1955-1956 academic year, I was brought on to wear one of his academic hats, the teaching of vertebrate embryology to pre-professional undergraduates, notably premedical and predental students. The embryology course was a spring semester course, but I was not given a free fall semester to get myself established, as would happen nowadays. The acting head of department designated me to teach laboratory sections of the elementary zoology course, just as I had done as a graduate student. During the ensuing next few years, I assumed a more typical teaching load of at least one major course and one or more graduate level courses each semester. My major courses were vertebrate embryology and comparative endocrinology, and my principal graduate or upper division level course was experimental embryology accompanied by neuroendocrinology, biology of sex, and so on. The load did not seem oppressive at the time but, in retrospect, I marvel that I was able to get a good start on my life's research activities under those circumstances.

When I arrived at the University of Arizona, none of my colleagues in zoology were engaged in laboratory bench work. Rather, the few active researchers were vertebrate zoologists who worked from field collections without the need of much bench space. Moreover, whatever equipment, materials, and supplies they needed were borrowed from the various courses they taught. I recall that only two members of my department had research grants and these were minimally funded. I was fortunate to have inherited the office of the recently retired department head and this space was large and convenient. Thus, I was able to start on my research program upon my arrival. Soon, each one of the faculty was to have good space because, within a few months, we moved into a new biological sciences building that was in the final stages of completion when I arrived in September 1956.

My move forward was much aided by the new availability of research support from the National Science Foundation (NSF), which had only recently instituted a class of low-level grants for young investigators. I applied for a Young Investigator Grant, and it was awarded in 1957. This was the first in a string of NSF grants that provided me with continuous research support for the next forty-two years until after I formally retired. I am grateful to the University of Arizona for allowing me to retain my space for minimal scholarly activity to this day. The onset and development of my investigative career occurred at a most propitious time, for these were years when federal funding, mostly from NSF or NIH, was generous. These were golden years, both for funding and for discovery, and the two fed on one another. In my view, this great progress in life sciences research was based heavily upon two post World War II events: the availability of radioactive isotopes and the development of electron microscopy, first transmission EM and much later, scanning EM. The former provided a potent means of cell component labeling that allowed many insights. With the latter, important ultrastructural and surface observations became possible. In these modern times, their uses have diminished as we have moved into the age of molecular genetics that require even newer technologies.

As we move along with the flow of the molecular events, as important as they are, I cannot help but look back with great regret, in part because much modern progress leaves in its wake the concept of the pursuit of knowledge for its own sake. Thus, the thread of investigation concerned with many old concepts that are still viable is often hung out to dry and bypassed, if not forgotten. I suppose that we must accept this chain of progress as being inevitable and normal. In fact, as I look back at my own research history, I recognize that many detours I took have left behind what I call "unfinished business." As I go on, I wish to consider some of these lesser-known or forgotten efforts, many embedded in the fabric of comparative biology that, in the long run, provided the basis from which our current knowledge has stemmed.

Aside from these more noble reasons for my writing this history, much of the motivation is personal. An important impetus comes from the need for mental exercise. As time passes, I note that more and more conversations with my aging friends are laced with descriptions of things or people because a term or name is not immediately retrievable from our relative banks of memory DNA. I hope that the mental exercise of writing will help preserve my capacity to recall and reason just as physical exercise has slowed the decline of my body. Another personal motivation is to acknowledge the indispensable contributions made to my research endeavors by many students and associates who have worked with me either in my own laboratory or elsewhere. The contribution of Japanese colleagues and partners is especially evident in the studies I describe.

Finally, this writing allows me an outlet to express a pride of accomplishment, not only from my part, but also from that of many others who rose from humble beginnings to make significant contributions to science and society. In my case, it is as the son of poor Italian immigrants who took advantage of what this country made possible. For others, it comes from deprivations due to race or culture. In all these cases, education was the key to success due to its availability and use. It is evident that the respective gene pools of all peoples are sufficiently strong to permit many to succeed if they are allowed access to education and freedom. There are many examples of this concept, but one stands out in the chronology of this narrative. I refer to the GI Bill that allowed veterans of WWII the resources and opportunities for higher education. Just think of the many GIs who would never have gone to college were it not for this support. Many went on to graduate school and to become renowned and accomplished scientists. In part, it was the efforts of these young people that fueled the burst of scientific progress that occurred in the three or four decades following the war. At present, we as a country are failing more and more to provide a strong education to the young. Even when good education is available, it has become a low priority, dampened by contemporary social pressures. One of my old French friends, Professor Louis Gallien, told us almost fifty years ago, after a visit to the United States when he was so impressed by our country, "*vous avez l'esprit des pioneers*" (you have the pioneer spirit) and that has given you your success. Perhaps we need a new infusion of pioneers to put us back on track.

Introduction

Some fifty plus years ago, at the start of my academic career, the various definitions of life science disciplines had different meanings than they do today. My PhD in Zoology meant that I was broadly grounded and that I had a degree of sophistication in a variety of subdisciplines. I thought of myself as an embryologist, but to the progressives, this title had a morphological connotation. Soon the more prestigious title "developmental biology" superseded embryology as more knowledge about mechanisms accrued. Similarly, the term "cytology" was superseded by cell biology. More specialized titles evolved as knowledge expanded. In my own case, I began to work with pigment cells and with pigmentation so, by the end of my career, I was considered to be a pigment cell biologist. I certainly did not intend to work with pigmentation for all of my life and, in fact, knew little about pigment cells when I started graduate work. I intended to be an experimental embryologist in the classic sense but, somewhere along the way, the vehicle pigmentation became the target. Over the years, I focused the tools of endocrinology, cell biology, and biochemistry to elucidate the mysteries of the pigment cell. Let me explain how this happened.

As I describe the research paths that I followed, I find it difficult to separate science from the joys experienced *en route*, so in order to accommodate both, the narrative proceeds in an anecdotal fashion. Justice is done to the science and at the same time is presented, I hope, in a manner readable and understandable to most who will read this lengthy story. At times I succumb to various diversions as I tell about travel, experiences, new friends made, and cultures encountered and somewhat understood. I wish to share with the reader things that we have learned with the hope that they enjoy hearing of them and vicariously enjoy the experience of research. What I cannot convey with words are the emotional benefits of the close and lasting friendships we have made with dear people from many countries. What also cannot be conveyed is the intense pleasure and satisfaction that I have derived from an active life of teaching and research.

Unfinished Business

1
Some Personal History

The Early Years

I am the son of Italian immigrants from Calabria, the province located at the tip of the toe of Italy. My parents met in the USA, but coincidentally both were from the same region, *Provincia di Reggio Calabria*. My father was born in the town of Delianuova on the slopes of the Aspromonte, the mountain range at the tip of the toe. My mother was born in Brancaleone Marina on the shores of the Ionian Sea. During WW I, my father was inducted into the Italian army. Taken prisoner by the Austrians near Trento while fighting in the Austrian campaign, he was interned as a prisoner of war in Bavaria. He spent fourteen months on a large farm. Just before the armistice was signed, he escaped and made his way back to Italy. It is not clear to me how he got home, but I do know that he was finally mustered out in Calabria.

In 1922, with the help of his elder brother, my Uncle Luigi (Louie) who had immigrated to the USA in 1903, my father entered through Ellis Island under the family name of Abagnara, the name that was also indicated on some of his military documents. Discussion with cousins of mine in Delianuova indicated that, indeed, Abagnara is the family name. Somewhere along the line, the surname was changed to Bagnara, probably by my uncle after he arrived in the USA. Thus, my father became Dominicantonio Bagnara after his arrival at Ellis Island. Since my father and my uncle were the last surviving Abagnara males, that name remains only on church and civil records and in cemeteries in Calabria.

After my uncle arrived in New York, for reasons unclear to me, he made his way to Rochester, where, during the twenty years before my father arrived, he prospered. In 1922, he was a union official in a clothing factory in Rochester, a position that helped him find a job for my father. It was here that my father spotted and took a fancy to an attractive young lady who came frequently to the factory to collect piecework that she took home to complete. Through the intervention of his much older brother, a traditional Calabrian courtship was begun. I do not know the details of this courtship since, at that time, I was, at best, a gleam in my father's eyes, but I do know that my maternal grandfather, Tomaso Zumbo, was a major player. The Zumbos lived in Victor, New York, a small town eighteen miles south of Rochester, and the "attractive young lady" was their eldest daughter, Fortunata, who would someday be my mother. It seems that the courtship involved my father going to call by taking a local trolley to the Zumbo home in Victor. There, the courtship involved my father sitting at one end of the sofa and my mother at the other end with Tomaso Zumbo holding down the middle. Anyway, the courtship was a success and here I am.

My father, Dominicantonio (Tony), and my mother, Fortunata (Fanny), taken in the mid-1950s, just a few years before her death.

The arrival of the Zumbos in America is interesting and probably typical of many Italian immigrant families. In this case, Tomaso and Rosa (Olimpia Bicchi) Zumbo were married and started a family in Italy. In about 1912, Tomaso left Brancaleone to find work in America. Rosa and three daughters, the eldest of whom was Fortunata, were left behind. For a long time Tomaso worked in the coal mines of West Virginia, but later he moved to Victor to join a colony of *"paisani"* from his hometown in Calabria, Ferruzzano. In Victor, he worked in a cider mill. The owner of the mill urged him to bring his family over from Italy. It was 1923 when Rosa, my mother, and her two sisters sailed to the USA. Eventually, this led to my becoming a first generation Italian-American, born in 1929, and a depression baby.

Our neighborhood in Rochester was ethnically diverse. There were many Italian families intermingled among immigrants whose origins were German, Ukrainian, Polish, Irish, Greek, and Jewish, mostly from Eastern Europe. They had in common the desire to succeed and a pride in who they were. These were strong families of industrious people. Despite the hard times of the Great Depression, they made do with whatever money they had and maintained their homes, yards, and themselves with pride. People on our street were neighborly. I remember well that during the warm weather of the summer, neighbors would come over after supper just to sit and talk in the front yard. Conversation had not yet become a lost art. We kids just loved to sit and listen to the adults tell stories.

Summer was also the time to prepare and preserve food for the winter. Our family canned vegetables, cherries, peaches, and pears. Tomatoes were preserved in prodigious amounts and in several forms, especially for use in making sauce (*salsa di pomodoro*), since pasta was consumed with meals several times a week. Pasta and meatballs were always a major part of Sunday dinner, and there was always enough for leftovers during the week.

Surprisingly, the ethnic groups on our street were not clannish; rather, they lived as individual families. However, related families did congregate on a regular basis. In our case, we often found ourselves at the home of Uncle Louie who lived only one house away from ours. Uncle Louie had married into a Sicilian family, the Anzalones. Aunt Mary had a myriad of Anzalone relatives, some of whom congregated at Uncle Louie's house almost every Sunday afternoon. The women talked and the men played Italian card games, especially *briscola*. The Anzalones were musical and, often, the men played together on the guitar, mandolin, or banjo. These are fond memories.

As a little boy, I was an avid reader, and I had the good fortune to have older girl cousins, my Uncle Louie's daughters, who took me to the public library every week. We, together with several other girls in the neighborhood, made the long trek to the Lincoln Library. In those days, walking was the mode of transportation. The walks were pleasant and, I think, served to reinforce the enthusiasm for reading. I read lots of books that were surely over my head, but this probably helped me in the long run as I learned from context. I have often wondered if my interest in science and nature came from this early reading but I really cannot say. Perhaps, it was just an inborn interest. In any case, for as long as I can remember, I always enjoyed being alone in nature.

Early on, I developed a fascination with birds. I attribute this interest to a strong exposure to migrating songbirds, especially the diversity of warblers that one finds in the East. I was fortunate to have a bedroom situated on the second floor of our house and that provided a close view of the large cherry tree in our neighbor's yard. It bloomed at the height of the spring migration and was a strong attraction to warblers in transit. Somehow, I had acquired a pair of inexpensive binoculars that facilitated identification of the little gems that flitted among the blossoms.

This was a time in my young life when I really needed a diversion from a harder part of life. I was only a preteen at the time, but I was much occupied with the serious business of housekeeping for my father and my sister, who was four plus years younger than I. These chores fell to me because my mother, who was afflicted with tuberculosis, was periodically confined to a TB sanatorium. One of her doctors was intrigued to learn that her young son was interested in birds. Consequently he gave her an elementary field guide to the birds that he had used as a boy. Of course, I was pleased and proud to receive this gift that allowed me to identify many of the birds I was seeing. Despite the fact that I had some serious responsibilities at this young age, I took it as a matter of course, because that is the way it was in those times.

My maternal aunts and uncles provided me with good examples of responsible behavior. When my grandmother arrived from Italy with my mother and her two sisters, after an eleven or more year separation from grandfather Tomaso, they made up for lost time. In short order, four more children were born, my two aunts and two uncles, the last of whom was only a year or so older than I. This was quite a burden for my grandmother because Tomaso died unexpectedly, and she was left to support four young children. My young aunts and uncles were faced with working to help meet expenses. During the summer, all four worked the sequential harvesting of vegetables and fruits on nearby farms, and thus they were able to buy clothes and books for the forthcoming school year. Every now and then, I went to Victor to participate in this summer work. I hated picking string beans, but stacking shocks of wheat was more to my liking, even for the meager pay of one dollar per day.

During those days when America was starting to emerge from the Great Depression, I never felt that we were really poor. I suppose that this was due to the fact that my father had steady employment. I recall that at one time, he brought home a salary of $18.00 per week. Our family was frugal, and we lived with a "depression mentality" that I think, in some ways, has remained with me until this day. Our yard was large enough for us to have a garden. I do not recall all of our crops, but we did grow tomatoes, romaine lettuce, and herbs, including basil and a plant that all our Italian immigrant friends called "*origano*." A much later search of my own revealed that this herb is actually summer savory and so, even in my waning years, I grow summer savory to season our tomato salad (*insalata di pomodoro*) or to put on grilled meat.

One of the onerous garden chores that we and many other Italian-American families undertook was preparing our fig trees for winter. The great majority of immigrants came from southern Italy and were passionate about fig trees. This was just part of our heritage. The climate of upstate New York was too cold in winter for fig trees, so we devised many ways to protect them from freezing. Usually, we made a board framework, bent the boughs low to the ground, and then covered the whole works with the fallen leaves of autumn. In the spring, we had to tear it all apart to expose the trees for summer growth.

It was emphasized to us during those depression years that we learn to save money, and I was taken to the bank to establish my own savings account. It was implicit that, in years to come, my savings were to provide for my education. It was a foregone conclusion that I was to go to college.

It was the aspiration of my parents, especially my mother, that I was to become a doctor. We Italian immigrants were much like other immigrant groups who had dreams that their sons would become doctors or lawyers and, thus, achieve success. It was all part of the American dream: become American citizens and then raise children who would achieve prosperity. My father became a naturalized citizen early on, and my mother followed suit sometime later. I remember that when I was a little boy, my mother enrolled in a citizenship preparation class given in the evening at a nearby elementary school.

Starting a childhood savings account was a long way from amassing enough money for college, but I was encouraged to save every penny that I could. At first, I did not do very well because it was difficult for me to have an after school job and also carry out my domestic responsibilities. Nevertheless, I did work in a grocery store for a short time, and when I turned sixteen, I got a summer job in a clothing factory. The big breakthrough came in the spring of 1947 when I was seventeen years old. I had heard that there were full-time jobs available as baggage and mail handlers for the New York Central Railroad (NYCRR). Although I was still a junior in high school, I successfully bid a job whose hours were from 6:00 p.m. until 2:00 a.m. This was a financial bonanza, for the job paid 97 ½ cents per hour. So, after a few hours rest after school, I walked the fifteen minutes to the depot and made the return trip home at 2:00. When the school year finished, I was often able to continue after 2:00 a.m. for several hours of overtime. The work was pretty hard, but I was a hefty kid who weighed about 220 lbs. The major part of the job was to haul baggage carts loaded with parcel post from the Central Post Office through tunnel to the far end of the train station where freight cars were loaded for destinations such as Buffalo, Cleveland, Detroit, Chicago, and St. Louis. These freight cars were called "the west cars" and were loaded by a crew that greeted the arrival of every baggage cart with a barrage of obscenities that added much to my education

The lack of sleep during the months at the start of the job was hard on me. I soon lost ten pounds and was often sleepy in class, especially after the lunch hour. However, I was saved by the start of summer vacation and, thus, I could sleep later in the morning. I really needed that extra sleep, because I often worked extra hours at time and a half. As the summer progressed, my savings increased remarkably, a fact that was aided by a pay raise to something close to $1.50 per hour, a pay rate that was pretty substantial for those times. As fall approached, I became a little apprehensive about how to budget my time for my senior year. A major concern was that I was a member of my high school football team and, while my work hours from 6:00 p.m. to 2:00 a.m. would give me time for football practice, the question of fatigue worried me. Fortunately, my job had improved, and I no longer had to haul carts from the post office to the depot. Instead, I worked trains; thus, when a train arrived, we were waiting on the platform with our loaded carts, and within a very few minutes, we transferred the loads to mail cars. Often, there was a sufficient period between trains to allow for a good rest or even a nap. By the end of the shift, the fatigue level was much lower than it had been. I endured through the football season and, when it was over, I bid a 4:00 p.m. to 12:00 a.m. job that provided considerable relief. By the time summer arrived, my load was even lighter because I put in some hours working in the station master's office doing clerical work.

During my senior school year, I made plans for college. One option was to accept a football scholarship at Niagara University and another was to try for acceptance at the University of Rochester. With the former, I only had the tuition covered, and Niagara certainly had less of a scholarly reputation. If the University of Rochester accepted me, I would have to pay a substantial tuition, but room and board were available at home. Acceptance was a problem since, in this post-WWII period of 1948, there were many returning veterans taking advantage of the GI Bill and admission pressure was very high. I had heard that only one out of eight applicants was admitted to the University of Rochester. Fortunately, I must have made a favorable impression during my interview with the director of admissions and I was accepted. The University of Rochester did not offer athletic scholarships, but I am sure that my being a football player was an important factor in my admission. Once accepted, I was able to matriculate in the fall of 1948 only because I had savings to cover tuition costs. At this point, I was exceedingly grateful to the NYCRR for many months of full-time employment; and soon, I had the opportunity to show my gratitude. As the fall semester of the University of Rochester was about to start, the head station master had a serious heart attack and everyone in the office moved up. Thus, there was an immediate need for a clerk to help the station master on the 11:00 p.m. to 7:00 a.m. shift. I was implored to fill in until a more permanent arrangement could be made. So, despite having a very heavy freshman load of science classes and, also, being a member of the football team, I filled in for about a month. I could never have made it were it not for the kindness of the night station master, John Bromley. The station was a busy place at midnight and for the next few hours as some of the New York Central's premier trains passed through. Among these were *The Twentieth Century Limited, The Commodore Vanderbilt, The Pacemaker, The Wolverine,* and so on. When the rush of traffic abated, John would often send me home saying that he could handle the rest. I remember well on the eve of our first football game, John sent me home just after midnight so that I could get some sleep. It was with some regret when I was finally able to terminate my job with the NYCRR and to focus, exclusively, on being a University of Rochester freshman.

The college years

What a welcome sense of relief it was for me to be free of my other responsibilities and solely concentrate on college life. My mother was still sick, but my father and teenage sister were doing fine without my help. I was happy with my classes, but I was especially happy to have a social life like any other guy my age. Perhaps, I was making up for lost time since I do not think that I was as good a student as I should have been. Nevertheless, I was able to prevail in the face of some tough first year courses.

The faculty at the University of Rochester was superb, and I was exposed to some fascinating and capable teachers who, in my case, "made a silk purse from a sow's ear." I could not have asked for a better education. My major was biology and most classes were relatively small allowing for good interaction with my teachers. They soon got to know me, and I to know them. All the senior faculty were eminent scholars with important research programs, and of course, their graduate students were outstanding. The very first course that I took, elementary biology, was taught by Professor Frederick Campion Steward, an Englishman who spoke in a captivating way that caused one to cling to every word. Professor Steward had just achieved great fame for he and his laboratory were able to cultivate a fully developed plant from a single fully differentiated cell. The premise was that every cell had its full complement of genes and that, if one could somehow activate the cell and get those genes to be expressed and to link in proper sequence, a fully developed plant would arise. He had found that if he took a small plug of tissue from a carrot root and exposed it to a factor (present in coconut milk) together with nutrients, individual cells from the carrot root would begin to form embryonic carrot plants that could be further cultured. I remember vividly walking through an underground tunnel connecting the biology building with an adjacent classroom building and seeing banks of racks containing roller flasks of plant tissue and growth factor that were rotating to expose the tissue to its culture medium. This pioneering work of Professor Steward's, done more than sixty years ago, was the forerunner of the mammalian stem cell cloning that is so prominent currently.

Aside from wonderful classes and teachers, I immersed myself in other aspects of college life. At the outset, of course, I was a member of the football team. Games were on Saturday afternoons and practice was every day. I loved our coach, Elmer Burnham, who was a wonderful person and a scholar of the game. He was one of the pioneers of the split-T offense and had come to the University of Rochester following WWII after he coached Purdue University to the Big Ten Championship. We were a pretty good team, and in my senior year, we won eight straight games before losing our last game of the season. My position was right tackle, and I played well enough to receive an inquiry from the Green Bay Packers and an accompanying questionnaire to complete. I ignored it. Not only did I have "other fish to fry" but, at 240 lbs., I knew I was too small for the pros. My roommate responded for me and filled in absurd data such as my time for the 100-yard dash as 9.9 seconds!

In those days, we played Hobart, Union, Rensselaer Poly Tech, University of Vermont, Amherst, University of Massachusetts, Wesleyan, and similar colleges. I loved our away games in New England at the height of fall colors. I have always been a jock, so sports were a dominant activity; however, I took part in many other social activities. One event I particularly enjoyed was an annual musical in which I had a small role. During my first two years, I had no interest in fraternity life, but some time in my junior year I was persuaded by some of my Jewish friends to join their fraternity, Kappa Nu. The fact that I was not Jewish made no difference to them. We were all friends and had many interests in common. I am sure that I was one of the first non-Jews to be a member of a Jewish fraternity in the USA, and it provoked a lot of humor. When I accepted their invitation to join, I did so with the provision that I would not have a bris!

My junior and senior years were exceptionally important because events therein shaped the direction my life would take. Like most biology majors, I took courses that were essentially premed courses required for admission to medical school. In my case, I was faced with a problem since, while my parents really expected me to go to medical school, I was far more interested in biology as a future. I was especially influenced in this direction by my close association with two of my professors. One was a young assistant professor, William B. Muchmore, and the other was Professor Johannes Holtfreter. The former, Bill Muchmore, was trained as an experimental embryologist but taught invertebrate zoology and was very much a naturalist.

Embryology was the domain of Professor Holtfreter who taught the basic course as well as experimental embryology. He and his students exerted a profound influence on my development as a scholar and future researcher and were instrumental in my decision to become an embryologist. Actually, Bill Muchmore was also influential in my choice for, although he and I shared a mutual interest in nature and natural history, Bill's personal research was in the realm of developmental biology. In fact, while he had taken his PhD at Washington University under the direction of the eminent Professor Viktor Hamburger, his dissertation research was also championed by Professor Holtfreter. Johannes Holtfreter and Viktor Hamburger were longtime friends since their graduate student days in Freiburg, Germany, where they were students of the legendary Hans Spemann, whose classic study on the organizer concept was published in 1924. In 1935, he and his student Hilde (Proksch) Mangold were awarded the Nobel Prize in Physiology for the work. That must have been a high-powered environment since the contemporary graduate students included Hamburger, Holtfreter, Proksch, and Paul Weiss, as well as Spemann's right hand man, Otto Mangold. About the time that I was an undergraduate, Hamburger, Holtfreter, and Weiss, then all professors at American universities, were regarded as the leading experimental embryologists in the world.

When I took Holtfreter's undergraduate embryology course, I had no idea how important a scientist he was, but what I did know was that he presented a fascinating course. I was particularly impressed by a mimeographed handout that he had devised. It depicted, with his drawings, a series of the classical experiments used to explain the organizer concept, embryonic induction, and the cascade of associated events. The last of the experiments were done in collaboration with his student, Chuang, in the mid 1930s just before his self-imposed exile from the increasingly powerful Nazis. In my own career, I used copies of this handout in every embryology course I taught.

Subsequently, I was even more overwhelmed by Holtfreter's course in experimental embryology. It was there that we learned the classical techniques developed by the German school of experimental embryology. We learned how to make tiny loops of baby's hair, to draw out glass needles, and to devise micropipettes. Most of our class experiments used amphibian embryos, especially those of the spotted salamander, *Ambystoma maculatum*. To insure survival of the embryos we operated on, we had to use sterile techniques. All of our tools and culture dishes had to be sterilized in boiling water. Around 1950 or so, penicillin and streptomycin had become available to add to our culture medium, and this enhanced the chances of survival of the operated embryos.

In addition to being a student in the course, I was very much involved in procuring the eggs or embryos for the experiments. I often went out with Bill Muchmore to a nearby preserve, Powder Mill Park, where there were many bogs and shallow ponds. Here, after an appropriate warm rain in the spring, male and female salamanders congregated in a breeding frenzy. Other amphibian species were similarly engaged and I learned much about them from Bill. There was also great birding here that Bill and I enjoyed.

There seemed to be no bounds to what could be learned, directly and indirectly, from the experimental embryology course. In many ways, we few undergraduates were trained as graduate students. We were each assigned seminar topics that we were to research and to discuss with the others. These seminar periods were usually held in our classroom, a lab reserved exclusively for our course. However, we were sometimes invited to Professor Holtfreter's home for seminar sessions. He lived in a single home with a small yard on Fitzhugh Street, in an old part of Rochester, near the center of town. That whole neighborhood was eliminated years later to make way for a renovated city center and hockey arena. The home was fascinating in that it was decorated with gorgeous paintings, murals, and other artwork all done by Holtfreter himself. Many were reminisces of a period during

A Christmas visit with Professor Johannes Holtfreter, my undergraduate inspiration at the University of Rochester. The photo was taken in the early 1990s. A sample of Professor Holtfretter's talent as an artist is displayed on the wall behind us.

WWII when he was detained by the British and exiled to the island of Bali. I have one amusing memory of one of these seminar visits. Professor Holtfreter knew that I had some knowledge of birds, so he asked me about a bird he had heard in his yard. He had not seen it that well, but he said its song was "tweet tweet" and he wished me to identify it. Of course, I could not, but sometime later, I realized he had heard a migrating white-throated sparrow.

In addition to the two formal courses that I had taken from Professor Holtfreter, I was enrolled in a special course to do individual research. The problem he assigned me related to his long interest in cell movements and the cell surface that he had begun working on just before he left Germany in the mid 1930s. When he was finally re-established after the war, first at McGill University and subsequently at the University of Rochester, he resumed work in this area. My project was related to a recently published paper by Katsuma Dan who was married to Jean Clark Dan, a Philadelphian, who I later learned was much revered by her Japanese colleagues.

His experiments concerned the stresses undergone by the cell surface of sea urchin eggs during the process of first cleavage. He had marked the surface with adhering particles of clay (kaolin) and then measured the changes in distance between particles as the egg divided. The changes in distance were presumably reflections of stretching of the surface. I was to essentially repeat Dan's experiments using the eggs of leopard frogs. I was too naive to comprehend the difficulty of the problem, and I believe that the same held for Holtfreter. Nevertheless, I was hot to go. Holtfreter had proposed that first I had to obtain naked frog's eggs, and then I was to prick the cell surface. There would be an exudate at the pricking site; when the cell membrane healed and sloughed the exudate, a discoloration scar would remain at the site of healing. I was then to measure the distance between the scars as cleavage proceeded. I was given a submaster key and a space of my own in the embryology lab, and away I went.

First, I learned to ovulate a female frog by pituitary implantation and then to fertilize the extruded eggs. Through discussions with the graduate students and by reading the literature, I learned to remove the egg jelly with the plant enzyme, papain, and to subsequently soften the fertilization membrane with thioglycolate. These procedures needed to be done relatively quickly because first cleavage occurs a couple of hours after fertilization. The next step was the most difficult and required great dexterity (which I had in those days). Under a dissecting microscope and with the use of fine watchmaker's forceps, I had to tear the softened fertilization membrane and release the egg. After many attempts, I finally succeeded. But, to my great dismay, the yolk-laden egg flattened like a pancake since it no longer had a stiff fertilization membrane to keep it round. Thus, the experiment was not possible, but I learned a lot and was ready to proceed to a new project! The induction of ovulation and subsequent fertilization of the extruded eggs were major techniques in my research career for as long as it lasted.

The next project that Professor Holtfreter proposed was exceedingly fascinating and held my interest for many years. It dealt with the phenomenon of embryonic induction and, in this case, the induction of the embryonic lens in overlying ectoderm by the optic vesicle which grows out to make contact with the ectoderm. This phenomenon had been much studied by the Germans and early day American embryologists. The question I was asked to investigate concerned a possible quantitative relationship between the inducer, the outgrowing optic vesicle, and the "reactor," that is, the overlying outer ectoderm. It was known that at the open neural plate stage, the prospective optic vesicle primordium was already determined, the right eye at the right anterior portion of the plate and the opposite for the left eye. Moreover, it had been shown that this determination was relatively loose so that, if only a part of an optic vesicle primordium were transplanted, it would develop into a complete optic vesicle, but smaller, and of a size proportional to that of the piece transplanted. In European salamander species (genus *Triturus*) which were used in these early experiments, the smaller optic vesicle induced the formation of a lens in the ectoderm it came to underlay; but no one had ever determined if the transplanted optic vesicle and the induced lens were of proportionate size. I was assigned the task of doing this using the American species, *A. maculatum*, the species of salamander available in our area. So, away I went using the techniques I learned in class. At the open neural plate stage, I would transplant different sized pieces of right or left prospective optic vesicle to the flank ectoderm. There each piece would heal, then roll up just as it would have done had it remained in place in the neural plate. The little vesicle so formed would be, at the same time, covered by ectoderm of the flank. In the embryos and the larvae that survived, a first measure of success was the development of a black spot or ring at the site of the operations showing that the pigment epithelium or the tapetum had differentiated. As to whether a lens was present had to wait for histological work.

I loved doing these experiments and, with free access to the building and the lab, I spent as much time as I could on the project. Since the stock embryos were maintained at near developmental arrest at the low temperatures of our refrigerator, the operating period was extended over several weeks. Nevertheless, since the embryos were collected in mid to late April and the end of my senior year was in June, prospects of finishing the project were negligible. I did not care and, apparently, Professor Holtfreter did not either, so I plugged away on the project at the same time that I carried on life as college student. I remember well one warm spring Saturday when my fraternity brothers and I played a softball game. When it finished, rather than going back to the dorm (there wasn't a bar near the campus), I happily went over to the lab to spend an hour or so with my embryos.

The impending end of the school year was not the only obstacle to the completion of the project; rather, it was the nature of the research project, itself. What I was trying to do was really a graduate level endeavor worthy of an MS or PhD degree because, aside from the embryonic operations, there was histological work to be done and an analysis of the data in light of a literature review. Fortunately, I had earlier taken a cytology course, the laboratory portion of which was microtechnique. Thus, I was able to fix, paraffin embed, section, and stain the operated embryos. I was at this point when the school year ended; nevertheless, I was still allowed to continue with all the privileges that I had before. In essence, I was a graduate student even if not in the formal sense. The graduate students essentially treated me as a colleague and were a tremendous help. In particular, I refer to Philip E. Townes (better known as Len) and Morris Smithberg, both PhD students of Professor Holtfreter. They each remained friends for many years of our respective professional lives. Len was especially helpful with my project and showed me how to make macrophotos of my operated larvae. His PhD dissertation was published as a classic paper (Townes and Holtfreter, 1955) that has been much cited. This research was a logical continuation of work started by Holtfreter in Germany at the time of the rise of Hitler and the Nazi party. It was at that time that Holtfreter saw the handwriting on the wall and escaped to England, ostensibly, on a lecture tour. He was there when the war broke out.

The story of Len Townes is an interesting one and is reminiscent of those times. His desire was to become a physician, but acceptance to medical school was difficult both at Harvard, his alma mater, and at Rochester because he was Jewish, since these schools operated on the quota system. While he was a graduate student, his application for admission to the University of Rochester medical school was rejected on a regular basis. However, in an act of cruel irony, he gained admission through the back door after he received his PhD. During the summer, the University of Rochester College of Medicine made a cadaver available to him so that he could learn gross anatomy. Thus, he could teach gross anatomy to the incoming class in the fall. At the same time, he was allowed to take the medical school courses and, eventually, he obtained his MD degree. Afterwards, he joined the medical school faculty as a pediatric geneticist.

During the few weeks after my graduation, I had time to examine the sections of my operated embryos. I was pleased to find that, indeed, the transplanted optic primordium did form optic cups and, in some cases, an early lens stage was present. At the same time, I learned from the literature that a transplanted *Ambystoma* optic vesicle cannot induce lens formation from ectoderm other than that immediately over the normally developing optic vesicle, in situ. It is a case of restricted competence in the Ambystomid family and is unlike the situation in the newt (Salamandrid) family where all the ectoderm is competent to respond to lens induction. So, then, what were the lenses doing in my experiments? There is no clear answer, but a possible explanation occurred to me as I looked at pertinent literature. These lenses were probably derived from "Wolffian regeneration," a term that the Germans applied to the regeneration of a lens from the dorsal iris. I was fascinated when I first read about it, and I remained so for many years. At any rate, this was the end of my undergraduate research project. It was not to be successful with this particular species. Nevertheless, it was not a futile endeavor, since I learned so much from all the efforts I made.

The summer vacation was now in full swing, so I turned in my keys and no longer went to the lab. I was now deeply immersed in the concepts of experimental embryology. I spent as much time as possible reading about the subject even beyond the guidance provided by our course text book, *Principles of Development* by Paul Weiss, a classic book that few people may now remember. In class, we had been introduced to a fantastic book, an opus magnum, *Biochemistry and Morphogenesis*, by Joseph Needham. I bought a copy for myself, and it became my bible for the summer. In fact, I was getting a jump-start on graduate school. During my senior year, I had decided that I wanted to further my education and to do research in embryology. It was obvious that I would need an assistantship to support myself, so I looked for universities that had both strong programs in zoology and the possibility of support. The Big Ten universities were an obvious choice. I applied to both Indiana University and The University of Iowa (then known as the State University of Iowa [SUI]). The latter was my first choice, not only because its catalog revealed the presence of at least two embryologists on the zoology faculty, but also because I had been enchanted by the movie "State Fair." The film contained such beautiful music and showed verdant tranquil scenes I found really attractive. I was excited to receive acceptance into the graduate program and an assistantship for the fall semester at the

University of Iowa. It meant both the start of a new life and the key to my future.

I wanted to hit the ground running and to emulate the graduate students I knew at the University of Rochester. In addition to Holtfreter's students, role models were Franklin Stahl, who later became a well-known pioneer in molecular biology at the University of Oregon, and Melvin Spiegel, who became a successful Ivy League cell biologist. It was sad to say good-bye to these friends with whom I played softball and interacted both socially and academically. My farewell to Professor Holtfreter turned out to be relatively temporary for I later saw him often when I returned to Rochester to visit family. Before I left, he asked that I extend his best wishes to his acquaintance, Professor Emil Witschi, a well-known embryologist at Iowa and one of the two names I had seen in the University of Iowa catalog. Doing so when I arrived in Iowa City turned out to be very important to my future.

2

Graduate School

Getting started

My new life in Iowa City was a large change, but altogether pleasant. I went from a city with a population of 250,000 to one of 25,000. In the former, the university was dwarfed by the number of people, the industries, and the cosmopolitan culture. In Iowa City, the university was the major industry and life centered around it. As with many Big Ten university cities, Iowa City was a university town. Stores and shops were intermingled with university buildings, although the heart of the campus, the Pentacrest, sat alone in its central dominance of Iowa City. In the center of the Pentacrest, elevated above the remaining four buildings, was the gold-domed former state capitol building. It is the historic symbol of the university and is affectionately called, "Old Cap."

My first remembrance of Iowa City was the dormitory, the Quadrangle, where I lived for a good part of my early years there. The Quad and other men's dormitories are situated on the west side of the Iowa River that essentially bisects the university. I was assigned a roommate, Chuck Wyllie, from Sigourney, Iowa. I suspect that we were selected to room together because we were both a little older, I, a graduate student, and Chuck, a veteran of the military. Actually, we had little in common, and since I spent most of my time in the zoology building, we really did not see much of one another. I was grateful to Chuck and his family for their invitation to spend

Thanksgiving weekend at their home in Sigourney. The visit gave me my first exposure to rural life in small town Iowa.

The first visit to my department in the zoology building sealed my fate. After checking in at the office, I decided that I should seek out Professor Emil Witschi to extend greetings from Professor Holtfreter. At that moment, I had no idea who would be my major professor, my thought being that this would be determined in due time. Professor Witschi greeted me warmly. As we talked, I realized that my salutation from Professor Holtfreter was my passport, and that Professor Witschi automatically accepted me as his graduate student. This meant that my teaching assistantship would be in his course, vertebrate embryology. With these two issues decided, my graduate experience was launched.

Professor Witschi was, by far, the most well known professor in the department. He was born in Bern, Switzerland, and had studied with one of the historic figures in embryology, Theodore Boveri, in Munich. While in Europe, he began a research career that eventually brought him to the faculty of the University of Iowa. He had a very impressive program at Iowa that was heavily supported by research grants. The principal theme of his program dealt with embryonic sex determination, but specific research topics took many directions from genetics to behavior. Laboratory space was substantial although spread out. The zoology building was a very old building which housed most of the department, and the Witschi lab had a large share. Associated with this building

Professor Emil Witschi, my dissertation director at the University of Iowa.

was an annex that included an animal facility primarily occupied by Witschi and those associates and students of his who worked with mammals and birds. During my stay at Iowa, activities in the annex were under the control of Sheldon (Shelly) Segal who had been an earlier PhD student of Professor Witschi and who had returned to work on several projects in conjunction with the Department of Urology. Just before I finished, Shelly left Iowa for a position with the Population Council at Rockefeller University. He became a very important person in the world of population issues and is often cited as the father of the contraceptive Norplant.

I recall that there were only two other graduate students in the Witschi group at that time. One, Don Johnson, spent most of his time in the annex on projects concerning mammalian reproduction. The other, Charles (Stretch) Sehe, was mostly involved with course work at the time. Stretch was a few years older than I and was doing his college work under the GI Bill. At the time, I knew that he had been in the Navy, but it was only recently that I learned of the depth of his involvement in WWII. It seems that Stretch had enlisted in the Navy before our entry into WWII. He was only seventeen years old. He was assigned to the battleship *USS Nevada* that was tied up next to the *USS Arizona* at Pearl Harbor on December 7, 1941. He escaped injury, but was involved in the horrendous clean up of bodies in the wake of the Japanese bombing. He was on the *Nevada* as it bombarded the shore at Utah Beach on the D-Day invasion. Later, he recorded the *Nevada's* shelling of the Mediterranean shore near Toulon as the Allies invaded France from the south. He was still aboard the *Nevada* when she sailed to participate in the Pacific campaign. One of the more notable battles he took part in was the taking of Mt. Surabashi on Iwo Jima.

The Witschi lab in the zoology building was managed by a wonderful person who became a dear friend. He was, using Chinese nomenclature of earlier times, Dr. Chih Yi Chang. He was called Chang by all who knew him. Chang and his wife, Florence (a microbiologist), were originally from Shanghai and escaped to the USA after the hostilities of WWII. I am not sure if Chang received a doctorate in China, but I strongly think so. At any rate, he obtained an American PhD from Professor Witschi at Iowa and stayed on as a research associate. In China, Chang had studied with Professor Tung who was the leader of the Chinese school of experimental embryology and who was probably classically trained in Germany.

Dr. Chih Yi Chang in the lab, about 1953.

Chang and I immediately hit it off because we both had learned classical experimental embryology and its salient techniques. This facility in working with embryos was exceedingly important at that moment because of a project that Witschi was carrying out through Chang. Actually, the research was in the new field of comparative endocrinology, but the approach entailed the surgical manipulation of frog embryos. When I arrived in the lab, only Chang could do these manipulations, but with the expertise I had developed as an undergraduate, I was able to pitch in and help immediately. It was tacitly assumed that I was a team player and would participate in the lab's research apart from my own work as a graduate student.

At this time in the early 1950s, the protein and peptide hormones of the anterior lobe of the pituitary gland were known and had been fairly well purified. These hormones were commercially available from the Armour Laboratories in Chicago and had been extracted mostly from bovine and ovine sources. It was assumed that, although a hormone was designated FSH (follicle cell stimulating hormone), or LH (luteinizing hormone), or TSH (thyroid stimulating hormone), or Prolactin, and so on, they were all, to a certain extent, intercontaminated. Witschi was interested in determining the degree of TSH contamination in each of these hormones, and to accomplish this, Chang was to employ a bioassay technique. More sophisticated chemical assays were not yet available. The bioassay was based upon the capacity of TSH to stimulate thyroid hormone (thyroxine) production in tadpoles, and in turn, thyroxine would induce the cascade of changes that occur in amphibian metamorphosis. The extrusion of the hind limbs is one of the first of the metamorphic events. Thus, the length of the outgrowing limb could be used as a measure of the amount of TSH in a sample to be assayed. The problem with this plan is that intact tadpoles produce their own TSH, and this endogenous TSH would need to be excluded. The obvious means for precluding endogenous TSH would be to use tadpoles with no pituitary of their own. This is where Chang (and I) came in. We operated on embryos to deprive them of their prospective anterior pituitary gland (adenohypophysis). The tadpole would grow, but never metamorphose; however they could be induced to metamorphose, or at least grow hind limbs, with injection of a TSH-containing fluid. Thus, they could be used in the bioassay of TSH.

The pituitary (hypophysis) of frogs develops from a placode (thickening) on the surface of the head area of a frog embryo between what will become the forebrain and the oral opening. With the use of a fine glass needle in one hand and a hair loop in the other, it was a relatively simple matter to extirpate the hypophysial placode. The lesion healed quickly, and the embryo, in due time, developed into an hypophysectomized tadpole. In order to provide enough of such tadpoles for the assay work, Chang or I would sit down at a dissecting microscope for hours and produce hundreds of operated embryos. For me, this was a task to undertake on a Saturday morning when I had no classes or teaching duties.

My first semester was, of course, busy. I had to learn my TA duties as well as to take a class load. One of the courses was physical biochemistry, an emerging subject of importance. I realized immediately that some knowledge of calculus would be helpful, not only for the course, but for my future development. Accordingly, I audited an undergraduate calculus class, even knowing that I could not give it my full attention. I remember it well because it is one of the most important things that I did in my life. It was taught by an older retired professor, Mr. Beck, who was kind to me even though I missed class from time to time. I gather that many of my classmates, including two coeds, were engineering students. One of these young women, Miss Schulze, did me a favor. I had left my textbook behind when I left class one day, and she kindly picked it up with the intention of returning it at the next class meeting. Since I missed the next class or two, she carried it around until I next came to class. Of course, I was grateful to her. After the first semester, I did not see her again until the school year was over.

During the summer of my first graduate year, I did not take any courses, but I remained in Iowa City and was involved in research activities. One day as I walked across the Pentacrest on one of the diagonal sidewalks, I heard my name called. I looked up and found that it was Lou Schulze from the calculus class walking on an approaching diagonal. We had not seen one another since the end of class but Lou, being the friendly person that she is, engaged me in lively and very pleasant conversation. I learned that she was from Ossian, a small town in northeast Iowa, but that she was spending the summer in Iowa City as student manager of the dining room of Currier Hall, an exceedingly large women's dormitory near the center of campus. I was smitten, and as we went our separate ways, I knew that I had to see her again. The occasion soon arose.

The University of Iowa and Iowa City are surprisingly active places during the summer, and one of the events is the annual performance of an opera. That summer, it was to be *La Boheme,* one of my favorite operas. Thanks to my father, when I was growing up, I was exposed to opera. He had, in his bachelor days, purchased a number of 78-rpm records of arias from Italian operas, especially those of Puccini and Verdi. I used to listen to them often, and I had become a steady listener of the Metropolitan Opera broadcasts on Saturday at noon sponsored by Texaco and hosted by Milton Cross. Nothing would have it then, other than I try to find Lou and ask her to attend *La Boheme* with me. Fortunately, I remembered that her last name was Schulze, so it wasn't too difficult to call Lou Schulze at Currier Hall and, to my profound pleasure, she accepted. Thus, our first date was going to a performance of Puccini's *La Boheme* at McBride Hall. I remember it as a wonderful performance, but even if it were not, I would not have recognized the fact from my place on cloud nine. We seemed to have had a great time despite the fact that we hardly knew one another. In fact, Lou didn't even recognize that the name, Bagnara, was Italian. When she spoke with her mother, she told her she had a date with a "fat Jewish guy!" Our first date was such a success that I asked her out again before we each returned home later in the summer, she to Ossian and I to Rochester. When we returned for the fall semester, we began seeing one another regularly.

At first, our dating was limited because we were both busy. Lou was a *rara avis* in those days as a woman chemical engineering student. Moreover, she was a serious and successful student. In addition, she was active in campus activities such as Student Council. That didn't leave much time for me. For my part, I had courses to take and to teach, together with being immersed in a very active laboratory. Gradually, we each realized that we were serious about one another, and after overcoming some early fears on Lou's mother's part, our courtship was happy and comfortable for the next two years. We spent holiday weekends at Lou's parents' home in Ossian, and I enjoyed being there. Northeast Iowa is a beautiful, somewhat hilly area, and I especially liked to drive through the back roads, often with Lou's parents, Hazel and Herman. We all loved and respected one another. I don't know that Lou and I ever became formally engaged. For sure, I could not afford an engagement ring, but it was definite that we would marry. Our two-year courtship was altogether wonderful, although we could not spend as much time together as we would have liked. The chemical engineering curriculum was taxing enough, but Lou added lots of extracurricular activities. One year she was chairperson of the Campus Chest drive. She was an officer in her dormitory both years, including the presidency her last year. These two years were important for me as I began to carry out the research that I could use for a thesis or dissertation.

Lou the coed at a football game.

Life as a graduate student

These first two years were important also as I became further acclimated to life as a graduate student. I no longer had time for the leisure activities that I had enjoyed at Rochester. I particularly missed going to the gym and to playing half-court basketball that had always been a passion. Fortunately, during the summer I was able to go to the Iowa Field House for pick-up games at least once a week. The competition was especially fierce because there were some talented players who came to do graduate work in physical education during the summer session. One of the players was exceptional, not only because he was at least six feet six inches tall, but also for the skill he displayed. By the end of the summer—this was 1953—I realized that this man was Harry Gallatin, a perennial NBA all-star for the New York Knicks and who was the leading rebounder in the NBA, even as an undersized center. In the 1953-1954 season he set the all-time record for rebounds in one game at thirty-three. He also obtained his master's degree in physical education from Iowa that same year. He later went on to coach in the NBA for New York and other teams. If he were playing today, he would be a multi-millionaire, and I doubt that he would have felt the need to pursue a master's degree.

One of the fascinating aspects of graduate school at Iowa was the interesting diversity among my fellow graduate students. First of all, there were a surprising number of Black graduate students. As I thought about it, a logical explanation emerged. This was a time of intense segregation in the USA that precluded admission of Black students to the major universities in the South. Thus, there were many small Black colleges, established to serve the many local Black people who sought a college education. As one would expect, the faculty of these institutions aspired to further their own educations at the masters and doctoral level. The University of Iowa and the other Big Ten universities, with their large and effective graduate programs, served as a mecca for such students. Their high degree of success served to attract even more Black students. The Black graduate students in the zoology department were really nice people with whom I had great rapport and friendship. One of my fellow students in a comparative physiology course stood out. I remember only that his surname was Ferguson and that he was on the faculty of a Black college somewhere in the South. He was very quiet with a modesty that was almost self-effacing. What impressed me most was his intelligence that seemed to far exceed my own.

My own office mate, in fact, was Black. F. Roy Hunter, who hailed from Richmond, Virginia, was bright, hard working, and had a great sense of humor. We became pretty good friends and socialized together with some of the other grad students, mostly white. F. Roy and I often played tennis together at courts that were shaded at one end by tall trees. He insisted on playing only from the shaded end, and when I asked why, he explained that even as dark as he was, he did not wish to become even darker from the sun. As proof, he showed me his left wrist, which displayed an obvious tan mark on skin that had been covered by his wristwatch.

A remarkable experience for me, a white person from the North, was a trip back from Rochester to Iowa City. I decided to take my 1939 Ford on a circuitous route by way of Richmond to pick up F. Roy. He and his wife lived with his mother-in-law in a rather affluent Black neighborhood. They were extremely hospitable and made me feel very much at home. One evening he invited me to a nearby nightclub. We had a fine time and everyone was very friendly despite my being the only white person there. The next day we left for Iowa City. That morning, his mother-in-law prepared a basket of fried chicken for the trip. As we drove west and on to West Virginia, I realized just how important that basket of food was, for we could not be served at any café or restaurant. As evening approached, we had to face the problem of where we would sleep. Finally, just after we crossed into southern Ohio from West Virginia we stopped for gas at some out of the way place. The owner decided to help by offering us a small bedroom attached to the gas station. The problem was that it had only a double bed, and I was a big person in those days. We had no recourse other than to take the room, and I had a good night's sleep. On the other hand, F. Roy had a bad night since he was small and was not heavy enough to counter the tilt that I imposed upon the mattress with my heavier body. In the end, we overcame all obstacles and made it to Iowa City.

Another novelty for me was the large number of graduate students from Arab countries. This, too, was typical of the times, because as these countries were becoming more affluent, they felt the urgency to educate further their more promising students. The USA was the country of choice to provide that education, because we were respected at many levels. In fact, this was a period of the great "brain drain" that was supplying us with many fine students. They came to be educated, but instead of returning home, they remained to take advantage of life in America. Moreover, there were positions available for the bright and talented. Nowadays, such foreign students often choose to go to other Western countries for their advanced education. As the Middle Eastern graduate students dwindled in number, they were replaced by Asian students. Many of these students are choosing to go home after their educations are complete, because the attractions of their respective home countries are greater for them than life is here.

One of my fellow zoology graduate students who became a close friend was Ahmed Zein-Eldin from Cairo, Egypt. Ahmed was a warm, outgoing guy whom everyone loved. Among his friends who also became my friends were Mohammed Ayub, an Egyptian from Alexandria, and Aziz Fouad from the Sudan. Ahmed was a year or so ahead of me, and after he received his PhD he began to be pressured by the Egyptian government to return home. His education had been paid for by the Egyptian government. He was in no hurry to return, however. He had recently been married to another one of our fellow graduate students, Zoula Pyle, from Des Moines. Moreover, Zoula had become pregnant soon after their marriage. Finally, Ahmed decided to return to Cairo, while Zoula and their infant son stayed with her parents in Des Moines. Ahmed hated it at the University of Cairo, and despite his being offered all kinds of concessions, he realized that he could not bring Zoula and the baby to live in Egypt. A real "cloak and dagger" episode ensued, and Ahmed was able to escape from Egypt by way of Lebanon. Ahmed was trained as a parasitologist, but upon his return to the USA, he went to medical school. He became a very successful psychiatrist in Galveston, Texas, where he and Zoula raised a wonderful family and fully enjoyed the American dream. Aziz Fouad finished his studies at SUI, and he and his wife, Lisa, moved on to Iowa State University where he obtained his PhD. He went on to a distinguished career in electrical engineering, mostly at Iowa State University. When Mohammed Ayub finished his PhD, he joined the faculty at Texas Tech University as professor of industrial engineering.

My personal research

As Chang began to carry out the injection of pituitary hormones into tadpoles, I took part as much as I could. We noticed that following the injection of prolactin, these tadpoles showed profound changes in their skin. The skin of an intact (control) tadpole is relatively loose due to its content of subcutaneous fluid. In marked contrast, that of hypophysectomized tadpoles is quite tight and is lacking much subcutaneous fluid. Prolactin-injected hypophysectomized tadpoles had very loose skin, even more so than intact controls. This effect seemed interesting so I thought I'd investigate it as a thesis problem in spite of the ribbing I took from F. Roy who accused me of trying to grow titties on tadpoles. As I began the research, I realized that it was going to be difficult to measure water content and subcutaneous dimensions. As a result, I was easily distracted by another change in the skin that seemed to be hormone-related and that could probably be measured more easily. It had to do with alterations of the integumental pigment cells. Thus, I began to focus on the pigment cell changes that were induced by all the hormone samples we had, and I let go of the subcutaneous water problem. In later years, I realized that this latter effect was a manifestation of the general role of prolactin in water balance physiology. Later, it became a much-studied phenomenon by the comparative endocrinologists.

The color of intact tadpoles is generally dark, mostly brown, black, or tan, due to the presence of many black or brown pigment cells, the melanophores or melanocytes. It was known for many years that these cells are controlled by a hormone produced by and released from the intermediate lobe of the pituitary. Thus, the hormone was first called "intermedin," but as more was learned about pituitary hormones, it became known as MSH or melanophore (-cyte) stimulating hormone. It is the presence of this hormone that mediates the dark color of most vertebrates. Its absence in hypophysectomized tadpoles causes them to be light-colored, in fact, the first biologists to successfully produce hypophysectomized tadpoles referred to them as being "albinistic." The pale color confirmed the successful hypophysectomy of the tadpoles we used in our experiments. As I began to look closely at our experimental tadpoles, I noted that those that were not injected (the controls) actually appeared silvery due to cells in the skin that were broadly spread out (the expanded or dispersed state) to cover the entire dorsal surface. On the other hand, in hypophysectomized tadpoles that had been injected with prolactin, these large silvery cells were no longer spread out over the surface and had, in fact, been reduced to small dots (contracted or aggregated). They seemed to appear punctate. It seemed that their silvery contents had aggregated toward the center of each pigment cell. The tadpoles had thus lost their silvery appearance. A look at the old literature explained that these silvery cells were pigment cells called guanophores, presumably because they were thought to contain purine crystals, notably guanine. It seemed then, that in intact tadpoles, the guanophores were more or less punctate whereas in hypophysectomized individuals, they were in a dispersed state, thus accounting for the silvery appearance. Obviously, some hormone from the pituitary was effecting this change from aggregated to punctate, and the immediate thought was that prolactin was the culprit. However, I soon found that injection of other of the pituitary hormones also caused guanophores to become punctate, so it appeared that I was dealing with intercontamination. To ascertain which pituitary hormone was responsible for the response was to become the subject of my dissertation research. I was enthusiastic to get started on the research. Professor Witschi shared my enthusiasm because he felt that the guanophore response might provide a new means to bioassay one of the pituitary hormones. He was in hard pursuit of methods to bioassay the gonadotrophic hormones, FSH, LH, and prolactin. Unfortunately, I had made this initial finding at the end of spring or early summer and, thus, there would be no leopard frog eggs available to produce embryos until fall, at the earliest.

As I began to think of other things, my impatience got the better of me, and I wondered if another kind of frog tadpole would be available to provide guanophores for my study. I knew of a site just south of Iowa City where bullfrogs were available; moreover, from a small sideline study that I had done, I knew that they bred in mid-July. Another possibility suddenly emerged in the form of a completely different type of anuran, the South African Clawed Toad, *Xenopus laevis*. *Xenopus* was, at the time, becoming popular in many laboratories in Europe and the United States and was initially used for pregnancy tests. The Witschi lab happened to have a breeding colony of these aquatic toads that was left over from Chang's dissertation research. At that very moment, since males and females were kept together in some concrete tanks, they were breeding freely, just as they would do in the wild. Every day, many individual fertilized eggs were deposited on the sides of the tank or on concrete blocks that were used to provide cover for the toads. I didn't know much about *Xenopus* or its embryos, but I thought that they might be usable for my work. Lou was going up to Ossian for a long weekend at home so this provided me some extra time to check out *Xenopus* embryos. Every day, I collected fresh eggs, providing a continuing supply of embryos at a proper stage for hypophysectomy.

Xenopus eggs are a good deal smaller than leopard frog eggs, and they have a faster developmental rate. Their embryos are quite delicate and their form is different from that of the frog eggs I was used to. Nevertheless, I was able to access the hypophysial primordium in its location just below the forebrain and the oral plate (or stomadeum) that would later become the mouth. However, a major problem was that it was extremely difficult to extirpate the primordium without incurring some damage to either the forebrain or the stomadeum. Neural tissue is extremely forgiving at this stage and damaged forebrain would easily repair itself but the buccal endodermal area is much less so. Thus, a large percentage of the operated embryos had malformations of the mouth and would not be able to feed as tadpoles. Over a period of four or five days, I did three or four hundred operations, and eventually, I ended up with about thirty good hypophysectomized tadpoles. This was hardly an adequate number of tadpoles necessary to begin my proposed experiments; moreover, it turned out that *Xenopus* larvae completely lacked guanophores on their dorsal surface. Thus, I was stymied from going forth with the guanophore experiments until fall when I could again ovulate leopard frogs and obtain the embryos I desired.

A consolation for all the work on *Xenopus* embryos was an esthetic one. The hypophysectomized tadpoles were beautiful. In the absence of a pituitary gland, the melanophores were fewer in number and in a completely aggregated state. This fact, together with the total absence of guanophores, rendered the larvae mostly transparent. In the head area, the brain and brain case were clearly visible. Similarly, major nerves such as the optic nerve and other of the cranial nerves stood out, uncovered by the usual melanophores. Even glands such as the thymus gland could be seen clearly. I was proud of what I had wrought for I knew that this was a unique accomplishment. With no immediate need for these beautiful larvae, I decided to just let them grow. Every now and then, when my ego needed a boost, I would go to the lab to admire them.

One evening after dinner, I was walking back to my office when I decided to go into the lab. The lights had been off for several hours. When I glanced at my aquarium full of hypophysectomized tadpoles, I noted that their tails were dark and almost black. I was a bit crestfallen by the thought these larvae were probably only partially hypophysectomized and were now producing and releasing some MSH from residual pars intermedia cells. Then, I began to wonder why only the tails were black so, a few minutes later, I went back to look at them again, and to my even greater surprise, the tails were clear and transparent. This called for a little thought and so it dawned upon me that the darkening of the tail melanophores might have been related to their having been in darkness before my entry in the lab and that the lights somehow caused the tail melanophores to re-aggregate their pigment granules. The obvious thing to do then was to place some of the tadpoles, both intact and hypophysectomized, in darkness for an hour and then bring them out in the light. Sure enough, both the intact and hypophysectomized larvae developed black tails in darkness, and upon their return to room illumination, the tails began to lighten, becoming completely clear in about five minutes. This called for further investigation, so over the next few days, I did a variety of experiments, the most telling ones being performed on isolated tails. The hypophysectomized larvae were too dear and too valuable to sacrifice, so the isolated tails that I used were taken from normal tadpoles. I found that isolated tails, in darkness for thirty minutes or so, turned black, just as they would have done had they still been attached. Similarly, they blanched after a five-minute exposure to light. This result suggested that the tail melanophores themselves were sensitive to changes in illumination, and thus I designed an experiment to test this idea. It turned out to be very simple. By placing a microscope bulb in a microscope, I focused a small beam of light on a portion of the tail fin that was well populated with melanophores and then placed the whole works in the dark. Thus, most of the tail fin was in darkness and only the melanophores exposed to the beam were illuminated. After about forty-five minutes, the tail was removed for examination. I found that those melanophores under the beam were punctate, whereas those that had been in the dark had dispersed pigment granules. The obvious conclusion was that the melanophores are indeed photosensitive and this, in turn, raised the question of the mechanism involved. Of course, I was intrigued by this phenomenon but I was still in my early phase of graduate school so I could not continue this train of investigation although I knew in which direction to proceed. I terminated the study temporarily with the hypothesis that the tail darkening was much like the process of vision and was based upon the generation of visual pigments in darkness and their destruction in the light. My thought was that the visual pigments were present on the surface of the pigment granules much as the visual pigments rhodopsin and porphyropsin are located on the rods and cones of the retina. I published these experiments and hypothesis about a year after I received my PhD, and just a few years ago, a younger colleague in the pigmentary world confirmed the presence of visual pigments in the tail melanophores of *Xenopus*.

The summer was over, and now full attention had to be directed toward course work, teaching, and my dissertation work. There were two seminar courses that I enjoyed and profited from immensely: the first was neuroembryology taught by Jerry J. Kollros whom I liked and greatly respected. Professor Kollros was a descendant of the Paul Weiss laboratory at the University of Chicago. Professor Weiss, a fellow graduate student of Johannes Holtfreter in the Spemann laboratory in Freiburg, brought that formidable background with him to the USA. He became a pioneer investigator of neural development and did seminal work on nerve growth and neural pattern formation. He had several outstanding students including the Nobel Laureate, Roger Sperry. It was wonderful to learn about all of this through the voice of Jerry Kollros. Jerry became a friend for life, and we kept in contact regularly until his death a few years ago. Another extremely important course that eventually had an important influence on my life and career was experimental embryology taught for six weeks by Professor Louis B. Gallien, a visitor of Professor Witschi, I refer to him as "Big Lou" for that is what he became to me in later years. Big Lou was a professor at the University of Paris and had the study of sex differentiation in common with Professor Witschi. Professor Gallien's course in classical experimental embryology included material that I already had been exposed to during my undergraduate days at the University of Rochester. In fact, in some areas I may have been more knowledgeable than Professor Gallien. He recognized this, and at the end of the course, he urged me to apply for a Fulbright award after I finished my PhD work. He told me he would arrange for me to come to his laboratory in Paris. I knew that such a possibility could not occur right away because I would first need to find a job. Eventually, I did eventually obtain a Fulbright award to go to Big Lou's lab and so another lifelong friendship began.

At the same time that these courses occupied my attention, I began to made great strides in my dissertation research. After having studied the literature about the chemical characteristics of the pituitary hormones, I began to use this knowledge to deduce which of the hormones might be responsible for the pigmentary phenomenon I had observed, namely the induction of contraction of guanophores. After an array of experiments, MSH emerged as the strongest candidate. This conclusion was based upon several facts: 1) Injection of any of the hormone samples that caused the guanophore response (aggregation or contraction) at the same time induced melanophore dispersion (expansion). Moreover, the degree of response, contraction, or expansion was always parallel. 2) Treatment of the various hormone preparations in ways known to destroy their known activity, such as heating at high temperatures, did not affect the guanophore contracting or melanophore expanding capacities of the respective hormone preparations. 3) Heating of the preparations in alkali at 100°C actually caused the preparations to induce an even more profound response by guanophores and melanophores. 4) Logically, it would seem that if MSH were a hormone that was used to bring about color change, especially to adapt to background, on dark backgrounds guanophores should be contracted, while melanophores should be dispersed. On white backgrounds, the opposite should occur.

With the logistics of my research in place, I was now able to have a realistic view for the finish of my dissertation. Lou was to graduate with her Bachelor of Science in Chemical Engineering in the spring of 1956, and I hoped to obtain my PhD degree at the same time. This meant that 1955-1956 would be really busy. After solid preliminary experiments were completed, a steady availability of hypophysectomized tadpoles was required for future experiments. In the fall and winter, embryos were derived by induced ovulation. As spring approached, I tried to get eggs however I could, even those laid by females in the wild. The leopard frog that I used, *Rana pipiens*, was as common in the Midwest as it was in my home territory near Rochester. For reasons that I cannot now recall, I returned home for a few days in the early spring of 1955, and while visiting my relatives in Victor, I took a walk in the woods near my grandmother's old house. I was greeted by an enormous chorus of frog calls of all types that emanated from an easily accessible marsh. In the marsh, there was a large number of clusters of frog eggs I could tell were laid only a few hours earlier as they had only just begun first cleavage. I collected some of those, put them on ice in a thermos jug, and took them back to Iowa to hypophysectomize. And, indeed, I did.

I digress to my personal life because, about this time, Lou and I decided that we needed to start our life together. We made plans to marry in September at the start of the fall semester. We knew that we wouldn't have much money, but love and passion trumped practicality. Of course, there were other obstacles, especially for me. Lou was well into her last year of chemical engineering and had a relatively easy year ahead. I, on the other hand, faced a series of hurdles if I were to finish my PhD work in time to graduate by June 1956 along with Lou. It was a rarity for a woman to graduate with a bachelor's degree in chemical engineering. To have her and her husband receive his PhD concurrently would put the icing on the cake. So I attacked the obstacles, one by one. First, I needed to pass my two foreign language requirements, French and German. I was not worried about the latter since I had studied German, both in high school and college. French was another matter since I had never had a course in it. Nevertheless, on the prescribed date later in the spring, I gave it a try, thinking that my having had two years of Latin in high school would help and be aided by my childhood exposure to an Italian dialect. I was in luck, and I passed with the minimal score. The German exam would have to wait until summer.

One of the high points of this spring of 1955 was an immense honor bestowed upon Lou in recognition of her extraordinary performance as a chemical engineering student. Not only was she unique in being a woman student in this field at Iowa, but she was the top of her class in chemical engineering. By chance, Tau Beta Pi, the engineering equivalent of Phi Beta Kappa, was having its annual convention at Iowa State University in Ames. Because women were not eligible for membership in Tau Beta Pi, Lou was named to receive a woman's badge and was invited to receive this award at the convention. Of course we drove the relatively short distance to Ames where Lou was provided lodging in the student union hotel facility. I was in luck because our friends Aziz and Lisa Fouad were living in Ames and kindly put me up. How proud I was of her the next day when she was awarded the badge at the Tau Beta Pi banquet. The symbol of membership in Tau Beta Pi is the "Bent," but this was not awarded to Lou because of her ineligibility; however, the woman's badge has a "Bent" embedded in it. This award probably no longer exists since women are now admitted to full membership in the engineering honor society.

With the French foreign language requirement taken care of, I was ready to tick off the next milepost in my graduate school journey. Before that, however, I was offered an unexpected opportunity. As with many students in my position, the possibility of spending time at the Marine Biological Laboratory in Woods Hole, Massachusetts, was a coveted objective. The embryology course that was offered was especially attractive because it was team taught by a group of the most respected developmental biologists of the time. The possibility for me to enroll in the course arose suddenly when Professor Witschi decided to carry out a summer research project on underwater hearing at the Oceanographic Institute, also in Woods Hole. He knew that I had no tuition money for even the first half of the embryology course, so he offered to hire me during the initial six-week session. My job was to harvest pituitary glands from fish and to preserve them in alcohol for transport to Iowa City. Of course, I accepted this unexpected windfall and was thus able to join the Iowa City entourage at Woods Hole. This was not going to be a total detour from my dissertation work, however, because I could use pituitaries from some of the piscine species I would collect. I planned to assay their ability to affect guanophores and melanophores of my hypophysectomized tadpoles.

As the summer began, I went east to Rochester carrying with me a large gallon-sized thermos that contained hypophysectomized tadpoles that were derived from the eggs I had collected there in the spring. After a quick hello to my family, I continued by train to Boston and to Woods Hole. As soon as possible after my arrival, I found the embryology teaching lab, and to my good fortune, the teaching assistant, Roger Milkman, was already there. This was well before the other students were to arrive, so Roger helped me find a large aquarium that we filled with fresh water to accommodate my hypophysectomized tadpoles. He, like almost every other zoologist in the world, had never seen anything like them. Here were these beautiful golden shimmering tadpoles. Of course, I had to explain all about them and what

I was planning to do with them. Roger was a graduate student himself and understood full well their importance to me. In keeping with his own good sense of humor, when the other students (and faculty) began to arrive, he announced with a straight face that these curious creatures were whale spermatozoa! Although our careers went in different directions, I kept in contact with Roger for many years thereafter.

The embryology course was fascinating. It was as successful as it was because one person was more of less in charge, despite the transitory participation of several eminent embryologists for various periods of time. The course organizer was J.P. (Trink) Trinkhaus from Yale University, and he kept things in order. Among the other faculty were Ed Zwilling (Connecticut), Nelson Spratt (Minnesota), Mac Edds (Brown), Meryl Rose (Illinois), Cliff Grobstein (NIH and, later, Stanford), and Clem Markert (Johns Hopkins). Except for my first exposure to some marine forms, much of the course content was familiar to me. There was a large carry-over from what I had learned at Rochester as an undergraduate and from my early graduate days at Iowa. Moreover, I had faithfully kept up with the current literature. Thus, I was an active participant in the discussions that were part of the course. Apparently, I made a good impression on the faculty because as the end of the six-week first session approached, I was asked to stay for the second session with all expenses paid. It was sad for me to turn down the offer, but Iowa City beckoned. First of all, I missed Lou terribly. Moreover, I needed to pass my foreign language requirement in German, and I had to find a job to earn a little extra money before our forthcoming marriage.

At some point during my Woods Hole tenure, I needed to fulfill my obligation to Professor Witschi. When the day came, I made my way to the MBL dock where I was confronted with an enormous pile of small sharks, better known as dogfish, mostly *Squalus acanthus* and *Mustelus laevis*. With a kind of X-acto knife in hand, I began cutting through the cartilaginous roof of the mouth to access the base of the brain to find the pituitary gland of each of these several hundred sharks. I don't remember how many vials of alcohol I had, but many were necessary to accommodate this large number of glands. It was an onerous task that took all day, and the only company that I had were the many sea gulls that gathered by my side in the hope of getting a bite to eat. Of course, I saved a few of these glands, as well as some from other unrelated species that were part of the catch. Those I took to the lab to test on my hypophysectomized tadpoles.

Sometime before leaving Iowa City for Woods Hole, I realized that the tadpoles I had hypophysectomized were not those of the leopard frog, *R. pipiens* that I had been using for my earlier experiments. It seems that the eggs I had collected in the marsh near my grandmother's home were wood frog eggs (*R. sylvatica*). Nevertheless, they were perfectly good tadpoles with beautiful guanophores that responded to the pituitary hormones just as did those of the leopard frog. At Woods Hole, I had no means available to attempt to purify whatever hormones were present. Thus, instead of having a medium I could inject into the tadpoles, I resorted to mincing up the pituitary glands and putting the macerated glands into small bowls with a few tadpoles. This proved to be remarkably effective, for the tadpoles displayed prominent melanophore expansion and guanophore contraction. With respect to the latter, the guanophores became reduced to small spots, and in a few days, even the spots disappeared. This profound reaction was actually due to a total loss of the guanine and other crystalline contents of the guanophores. Studies of this phenomenon were important parts of my research in subsequent years.

I was not alone in bidding fond farewell to Woods Hole, for I was joined by Chang, his wife, Florence, and their two young boys, Robert and Howard. They had driven to Woods Hole from Iowa City and were returning early, just as I. Early in the spring, Chang had purchased a used car, and I was very much involved in teaching him to drive. At any rate, I hitched a ride with the Chang family as far as Rochester and then took the train on to Iowa City.

The Chang family's stay in Woods Hole was shortened because of their need to be in Iowa to file a notice of intent. These were difficult times in Sino-American relations in terms of the large question of repatriation. There were many American citizens in China that the Communist regime held and prevented from returning to the USA. At the same time, there were many Chinese citizens, students, scholars, and others, that the Chinese government felt we would not release. My understanding was that a repatriation agreement between the USA and China was well underway, and that in anticipation of the signing of the agreement, the American government had set a deadline (I believe it was August 14, 1955), at which time Chinese scholars needed to affirm their intent either to remain in the USA or to return to China. Chang and Florence had decided to return to China. I know that it was a difficult decision for them, but I understood full well the reasons for this decision. First of all, Chang's position had no security since he was paid from Emil Witschi's research grants. Professor Witschi was no "spring chicken," and when he was gone, so would be Chang's position. Chances of Chang's getting a position of his own would be very slim. Moreover, Chang remembered with what reverence a professor was regarded in China, and he aspired to such a position. Another important consideration was Florence's aging parents. So, the Changs had this dilemma hanging over their heads. Although the decision for the Changs was imminent, their actual departure for China would not take place until the fall of 1956, after Lou and I had left Iowa City. I, too, faced several hurdles as we headed west.

Within a few days of my return to Iowa City, I made an appointment to fulfill my German foreign language requirement. So, one mid-morning I went to the German department where the examination was administered by a pleasant young man, probably an instructor. I passed the exam quickly and went on to my next objective-namely to get a job for a few weeks. At that time there was major construction on US Highway 6 that went through Iowa City (Burlington Street), and I knew that jobs were available. Off I went to the construction site. When I arrived at a corner where there were some workers hard at it, I asked where I could find the foreman. They pointed to a man at the next corner in conversation with another man. When I approached, I realized that the foreman was talking with the very person who had just tested my German comprehension. He, too, was looking for summer employment and for the next few weeks we worked side by side.

In the meantime, Lou had been at work in Cedar Rapids. From the start of the summer, she worked as a chemical engineer at Cryovac, a company that made shrink-wrap. She had my car (Hazel, after Lou's mother) for the summer and shared an apartment in Cedar Rapids with an SUI friend. We had already found an apartment in Iowa City for the fall. It was conveniently located across the street from the zoology building and seems to have been sequentially rented over the years by a string of zoology graduate students. I don't remember when we took occupancy, but probably about

September 1. I know that I stayed there until our wedding on September 10. It was a fourth-floor apartment accessed by a narrow stairs that I got to know rather well as I carried up our meager belongings. A major chore was to haul up a large steamer trunk that I had brought from Rochester. It was the very trunk that had come from Italy when my grandmother, my mother, and her two sisters immigrated. It became a major storage place for us and continues that function even today.

We were married on September 10, 1955, in Danforth Chapel, a beautiful, small, interdenominational chapel on the banks of the Iowa River on the University of Iowa campus. We were married by the minister of students at the Wesley Foundation. I, of course, was raised as a Catholic, but by the time I was well into my education, I had no formal religion and could care less about who married us. Lou's father was a non-practicing Catholic, and her mother was a Protestant, primarily a Methodist. Thus, Lou was essentially a Methodist. I never thought very much in religious terms about our marriage ceremony, and over the subsequent years, we just drifted away from organized religion. During our courtship we did go from time to time to a Methodist church kitty-corner from the zoology building primarily to listen to the sermons (lectures) of the pastor, Dr. Dunnington, who spoke little of religion and much about world events.

The wedding ceremony was lovely. Lou was a beautiful bride. Her sister, Dorothy Johnson, was matron of honor, and Dorothy's husband, Eldon, was best man. Mohammed Ayub, a good friend and engineering graduate student, served as usher. Herman and Hazel, Lou's parents, represented her side of the family, but my family was not represented. My mother was too weak to make the trip, and besides, she was a bit dismayed that I was not being married as a Catholic. Under the circumstances, neither my father nor sister would make the long trip to Iowa.

As we were growing up, my sister and I had our First Communions and Confirmations, but we were not regular churchgoers. My parents never went to church except for weddings and funerals. However, when my mother's tuberculosis worsened, she became more religious. The disease got the better of her, and in the fall of 1957, a bit short of her fiftieth birthday, she passed on. If she could have held on for another year or so, she would have probably escaped because streptomycin, an exceedingly effective antibiotic against tuberculosis, was just becoming available. After her death, my father began to go to church regularly, and we respected that when he came to visit us in our Tucson home.

After our wedding reception later in the afternoon, we drove off to Dubuque to spend the first night of our honeymoon at the Hotel Julian. I'm sure we had the look of honeymooners about us as we arrived. Lou had on her beautiful going-away dress and hat, and as I nervously went to register us, she hid behind a potted palm tree! Dubuque is a river town of great historical significance, and the Hotel Julian is part of that history. We didn't have much time to spend in Dubuque being as penurious as we were. Instead, we drove on to Madison, Wisconsin, to spend two nights with former graduate student friends, Jerry and Hildegard Sherman. We had a three-day honeymoon.

Our apartment in Iowa City was pretty simple. The main room was multipurpose and served as living room and dining room. The kitchen was in a short, narrow corridor with a stove on one side and a sink on the other. There was not room for the refrigerator, so it occupied a spot in the bedroom. We did have good windows to the west that gave us good light, but also let in the hot sun in the summer. It was plenty modest, but we paid only $47.50 per month rent. We really couldn't afford more given that our primary income was my assistantship of about $180 per month. Lou's salary that she earned from working in the Department of Chemical Engineering was meager. We didn't have much to live on, but we made do. We ate frugally—this translates into high calorie foods. A big item in our diet was baked beans and pork jowls—tasty and cheap! We were too proud to ask for help, which much later led to Lou's dad remarking, "You wouldn't ask a guy for a five."

The school year that was upon us was especially busy for me, but less so for Lou who had only a few important courses to take. We acquired new friends in the lab with the arrival of Giovanni Chieffi from Naples, Italy. He was spending two years in the USA and had already spent a year in Detroit where he had done some work in parasitology at Wayne State University. Giovanni was a couple of years older than I and had already completed both an MD and a PhD. His association with medicine was for practical reasons, but he was primarily interested in research. It was for the latter reason that he had come to work with Emil Witschi in Iowa City. Giovanni arrived in Detroit with his wife, Annamaria, and their infant daughter, Matilde, and while they were in Detroit, their first son, Lorenzo, was born. Giovanni and I became friends immediately. Over these subsequent fifty-six years, Lou and I have become dearest friends with the Chieffi family. During this long period, we have seen them often in Naples.

This last and final year of my graduate studies differed from the previous three, mainly because I had no formal class work. Aside from my own teaching time as a graduate assistant, I was much occupied with my own research work. A slight diversion from my dissertation work was offered by a little personal project that derived from observations made while looking for leopard frogs and their eggs. During my first year at Iowa, I learned of some ponds south of Iowa City that were essentially oxbows of the Iowa River. While "playing naturalist" in the area, I found that there were salamanders present, *Ambystoma jeffersonianum*, but no leopard frogs. Instead, the ponds were populated by bullfrogs, *R. catesbeiana*. Observations and measurements I made over the next two years showed that these bullfrogs had a one-year life cycle, instead of the two- to three-year cycle typical of bullfrogs in other regions of the country. In other words, these frogs went from eggs to tadpoles to froglets within the space of one year. I discussed these observations with Jerry Kollros, and he urged that I write a short paper describing what I had found. He suggested that I present the work at the annual meeting of the Iowa Academy of Sciences soon to convene at Grinnell College. With the important help of Lou, who typed the manuscript and helped draft an important graph, I submitted the manuscript for publication with Jerry as co-author. I did so because of the encouragement that Jerry gave me and because the staging series that I utilized was that for leopard frogs, published by Cecil Taylor and Jerry Kollros during Jerry's graduate days at the University of Chicago. Thus, my first scientific paper and presentation were "Abbreviated larval period of *Rana catesbeiana* in Iowa," published in 1956 in the Proceedings of the Iowa Academy of Sciences.

With this project completed, my full attention was directed toward writing my dissertation. The plan was still for me to finish the dissertation and to defend it in time for graduation in the spring at the same time that Lou would receive her Bachelor of Science in Chemical Engineering degree. Lou helped by typing early drafts of the manuscript and by offering editorial suggestions. Unfortunately, this dream was thwarted by Emil Witschi. First of all, he took his time in reading the manuscript, and second, he could not believe some of the quantitative results I had obtained. I had found that the minimal effective dose capable of causing melanophore expansion and guanophore contraction was 0.0001 μg of MSH per 0.05 ml of MSH solution injected into hypophysectomized tadpoles. I, too, realized that this miniscule dosage entailed precious few molecules of hormone, but the results were consistent, so I knew that they were correct. In order to convince Professor Witschi, Chang prepared an array of unlabeled, numbered vials, each containing different concentrations of MSH. I was to inject samples from each of the vials and then report back what the relative concentration of MSH was for each vial. When I did so correctly, Professor Witschi was convinced, and my dissertation was accepted. Unfortunately, these machinations took too much time for me to meet deadlines for spring graduation. Thus, Lou graduated in June, while I had to wait until August.

Another preoccupation of the spring semester was to prepare for the next stage of life. I planned to do post-doctoral work to better prepare for future employment, and I wished to do that work in a lab that was interested in pure developmental biology. In fact, I had already chosen a person with whom to work, Dr. James D. Ebert of Indiana University. His lab was studying mechanisms that controlled specific organ size. In other words, what makes a specific organ, such as heart, liver, kidneys, stop growing when they reach a given size? I wrote to Dr. Ebert to tell him of my interest. He quickly replied by first encouraging me, and at the same time, telling me that he had just been appointed Director of the Carnegie Laboratory of Embryology in Baltimore. If I were to go to work with him, it would be in Baltimore. With his advice, I prepared an application for a US Public Health post-doctoral fellowship with the plan of going to Baltimore. This was fine with Lou, for she knew of Davison Chemical Company in Baltimore, a division of W. R. Grace and Company, as was Cryovac, her summer employer. When contact was made with Davison, they expressed great interest in Lou and requested an interview in Baltimore. The amount they offered to purchase airfare for Lou was sufficient to cover both our costs if we went by car.

There was no way that we could make the trip in Hazel, our 1947 Plymouth coupe, so we thought of borrowing Herman and Hazel's relatively new Studebaker. Lou's mother came up with a better idea. Ossian

High School was abolishing their drivers' training program and was going to sell their 1955 Chevrolet for $1100. So, we dipped into the meager savings that we had and supplemented that with a loan from Lou's parents to buy the car. It became Hazel No. 2 and she served us well for many years.

Off we went to Baltimore in our year-old 1955 Chevy. This was great, because Lou could have her job interview, and both of us would be able to see Baltimore and to meet Jim Ebert, my prospective post-doctoral mentor. It was a wonderful trip. We liked Baltimore, Lou's interview went fabulously, and Jim Ebert was impressive. He even called the fellowship office at NIH and was informed that my application was a strong one. We returned to Iowa City in high spirits that were further buoyed when Lou learned that Davison had offered her a job at a substantial starting salary. It appeared that between Lou's salary and my post-doctoral stipend, we were going to Baltimore and would live in high style.

Now that we had a reliable car, our trips to Cedar Rapids were a lot easier. We didn't need to go that often, but we did enjoy visiting Dorothy and Eldon. During our courtship days, they invited us often for dinner in their small apartment. They were already making plans for their house that was being built in a new development. We enjoyed the building project vicariously and were happy to visit after the house was finished. I made one such trip with Giovanni, because Eldon was very much interested in meeting him. Eldon had served in the entire Italian campaign of WWII. He was the map carrier for his company, and thus had an excellent knowledge of the country including Naples, where he had experienced the eruption of Mt. Vesuvius that took place at that time. From Sicily, he had landed just south of Salerno and then again at Anzio, just south of Rome, in one of the bloody battles of the campaign. He wanted to share these experiences with Giovanni who was a teenager in Naples at the time.

Dorothy and Eldon's house came none too soon, because in the early fall of 1955, Dorothy became pregnant with their first child. The baby was due in June, and this was a great source of amusement for us because she must have been conceived about the time that we were married. We liked to tease Dorothy and Eldon by suggesting that the baby must have been conceived on our honeymoon! After the baby, Sara, was born, we had even more reason to visit them in Cedar Rapids. Sara was a beautiful baby and was a lovely blue-eyed child. She became a wonderful young woman and excelled at Washington University in St. Louis, a sister school of my alma mater, the University of Rochester. Sara went on to law school at the University of Chicago and became a successful lawyer in Chicago. We loved her and could not have been more proud if she were our own daughter. After a relatively short but successful tenure with a well-respected Chicago law firm, Sara chose academia. She became special assistant to the chancellor of Washington University, and later was an associate dean of Liberal Arts. Life dealt her, and all of us who loved her, a terrible blow in the spring of 2004 when she was taken by a kind of pancreatic cancer. She was not quite forty-eight years old.

The year 1956 was not an ideal time to find a university position. First of all, it was still too soon after the end of WWII for there to have been sufficient financial recovery to reach the educational needs of the country. Moreover, the GI Bill had allowed for an increased number of PhDs who were looking for jobs. Thus, when I started to look to the future, I thought it best to hedge my bets by following up on several job possibilities. One good prospect was at St. Louis University, a Jesuit institution, where an embryologist was being sought. They must have found my credentials attractive for I was invited to St. Louis for an interview. It went very well, and I thought I had a good chance for an offer until late in my interview with their Dean of Liberal Arts, Father Marchetti. He was a very pleasant, massive man, an exact image of his more famous nephew, Gino Marchetti, the star tackle for the world champion Baltimore Colts. When our conversation turned to my religious persuasion, I was frank and told him of my Catholic upbringing, but that I no longer practiced the faith. Immediately, I sensed a change in the tenor of the interview, and I knew then that I would not be offered the position. Had I been Jewish or of some other faith, I would surely have received an offer, but a Catholic turncoat—no way!

The rejection from St. Louis did not bother me much, because we were set on going to Baltimore. As time wore on, however, I began to be concerned as to why there had been no word from NIH. When I telephoned NIH, I was told that my application looked very promising and that I should not be impatient. Then, one day, I received a call from Albert R. Mead, acting head of the Department of Zoology at the University of Arizona, offering me a position. Early on, I had made a token application for a position as an embryologist at the University of Arizona. However, I didn't really want to go to Arizona because of my own arrogant bias that it would be an academic wasteland. I stalled Al Mead as best I could, thinking that I would hear from NIH at any moment. After about a week, Al called again and upped the ante (to $4800 for nine months). I asked for forty-eight hours, during which time I again called NIH. Again they gave me an enthusiastic response, but when I asked, "Do I or do I not have the fellowship?" they indicated that they could not give me a definitive response. When Al Mead called for a final answer, I accepted the offer at the University of Arizona. In retrospect, had that fateful day been now, I would have rejected the offer with the full knowledge that Lou would have a job in Baltimore with sufficient salary to take care of us, even without the fellowship. However, in 1956, things were different, and the husband was supposed to be the breadwinner. Now that women have achieved at least breadwinner status, it is for the better. It will be even better when women achieve equal salary status.

There was much to do between my graduating and our departure from Iowa City. Lou had again worked for Cryovac during the summer of 1956, and she planned to continue work until the last minute. Much of the moving fell on my shoulders, literally and figuratively. I had some heavy items to haul down four flights, my mother's trunk being the heaviest and most awkward. Working side-by-side with me was Hazel, who had come down from Ossian to give us a hand. Finally, departure day arrived, and as we bid fond farewell to four wonderful years, we headed east. With my mother's precarious health, we felt it important to go to Arizona by way of Rochester, NY. The trip west was long and most interesting for neither of us had been through the South or the Southwestern states. We arrived in Tucson on September 3, 1956, at the end of the monsoon season, and yes, we were greeted by a thunderstorm.

3

The Start of a Professional Career

Settling down in Tucson

After finding a place to live, a single unit in a small complex about two miles from the University of Arizona campus, we each turned to matters of employment. I had my position on the lowest rung of the academic ladder, Instructor in the Department of Zoology in the College of Liberal Arts. I suspect that the State of Arizona was still using this lowly rank, because the starting salary for university faculty was so low. In my case it was $4800 for the academic year of nine months, up from the original salary of $4200 that Al Mead first offered me. My next year's contract was at the rank of Assistant Professor with a modest increase in salary, just enough to keep me from leaving. Lou had no job when we arrived, but was fortunate enough to find a position as a chemical engineer in short order. In 1956, southern Arizona was not exactly a haven for industries employing chemical engineers, but Infilco, Inc., a water and waste treatment manufacturer, had moved to Tucson from Chicago several years earlier. Lou dressed up in her finery and took herself to the plant, where she had a successful interview and was hired as a project engineer. She enjoyed her job, was respected, and made lifetime friends.

I joined the zoology faculty at a time of great change. The former head of the Zoology Department, George Caldwell, had just retired a few months earlier and was replaced by an interim head, Al Mead, who hired me. In addition to being head, Dr. Caldwell taught embryology and physiology.

My job was to replace him as the embryologist, and another newly hired faculty member, William J. McCauley, took on the task of teaching physiology. Bill McCauley was a native Arizonan, and we naturally became friends as he helped me get acquainted with the state. I inherited George Caldwell's office which, having served the head, was large in comparison with space the rest of the faculty had. This was of no particular significance since our department was to move into a new building in the spring. Everyone would have more than adequate space, especially compared with what was then available in the Speech Building where our department was situated. Not only was the Speech Building small, but it housed the Departments of Speech, Zoology, and Microbiology, as well as the College of Education.

In 1956, the particular charge of departments in the College of Liberal Arts was teaching. Thus, despite the fact that my course, vertebrate embryology, was scheduled for the spring semester, I was immediately put to work teaching a couple of laboratory sections of the elementary zoology course in the fall semester. This course was taught by Professor William H. Brown, an older gentleman who had been in the military during WWII. Bill became a good friend and was quite an interesting character. As an example, when I first introduced myself to him, out of deference to his age and position, I addressed him as Professor Brown. His response was, "Don't call me Professor; a professor is a guy who plays the piano in a whorehouse!" Bill and I would often go for coffee in mid-morning and *en route* he would introduce me to many university people. Often, on Friday night, Bill would invite me to his home, ostensibly to plan the teaching of the upcoming laboratory periods, but practically to drink a little Scotch whiskey together. Bill had been a widower, but a few years earlier had remarried a wonderful woman, Elena Hornaday. Elena was from Arlington, Texas, then a quite distinct small city between Dallas and Ft. Worth. She owned and managed a highly successful Mexican restaurant, La Tapatia, which drew many customers from the two adjacent metropoli. The restaurant seemed to function well even in her absence, as Elena lived primarily in Tucson after she married Bill.

When we arrived in Tucson and I started work, I went with the flow of my personal nature and "hit the ground running." My first semester teaching load was easy since I came from Iowa with good teaching experience in general zoology. When I assisted in the general zoology labs in Iowa

City, among my students were some of the well known football players on the stellar teams that Forest Evashevski had recruited for the Hawkeyes in those days. The same was true at the University of Arizona. Three of my student players at the University of Arizona I remember quite well: John Malekas, Bob Whitlow, and Jack Young. Malekas and Whitlow went on to play with distinction for the Chicago Bears. I liked them and found them amusing, but not so for Lou. In those days, since we had only one car, Lou would drop me off at the Speech Building *en route* to her job at Infilco. Often, when she picked me up at the end of the day, she let me know that my football-playing students had "wolf-whistled" her on their way to my class. Once, at the same time that she picked me up, one of our common birds, the curved bill thrasher, flew by uttering its characteristic whistle. When I identified the call, Lou admitted with some embarrassment that what she thought was a wolf whistle from my students was actually that of a curved bill thrasher!

The space that I had inherited from George Caldwell was sufficiently equipped for my immediate needs. It had some surface space, running water, and gas, so I got started with my research immediately. I used leopard frogs from the zoology course to obtain embryos, and I made the appropriate tools for microsurgery with which to operate. Fortunately, I was able to purchase Swiss watchmakers' forceps (Dumont fils #5) from a local jewelers' supply company. Watchmakers' forceps were essential for removal of the jelly and membrane that encloses the embryos. I was soon at work depriving these embryos of their prospective pituitary glands, so I could continue my work on the control of the bright colored pigment cells. It would take about six weeks for the tadpoles to grow large enough to work on. I used this period to search for funds to carry out the research.

Taking Root in Academia

The late 1950s and 1960s were a great time for a young investigator to be starting out, for grant funds were gradually becoming available. I took advantage of a program that NSF had established for young people just getting started. This series of young investigator awards provided small grants that required simple application procedures and quick funding. So, in just a few months after our arrival in Tucson, I submitted a grant application to NSF and was fortunate enough in the spring of 1957 to be awarded a two-year award of $7,000. This marked the beginning of a continuous series of National Science Foundation (NSF) grants that lasted through the next forty years. By taking advantage of my relatively small teaching encumbrance, I was soon able to begin to prepare my dissertation for publication. It went a little slower than I would have liked, but in those days, so long ago, I was a neophyte at scientific writing. At the same time, I wrote a smaller manuscript describing my experiments on the tail-darkening reaction in *Xenopus*. This took a shorter time to complete, both because I had prepared all the photos and illustrations I needed in Iowa City, and because the project was smaller. Thus, the latter was published in 1957 and my dissertation appeared a year later in the *Journal of Experimental Zoology*. I had chosen to submit the tail-darkening manuscript to the Proceedings of the Society for Experimental Biology and Medicine because I had become acquainted with the head of the Department of Biochemistry, Arthur Kemmerer, who was a member of the Society. A prerequisite for publication of an article in their journal was an introduction by a member. Art Kemmerer was a lovely man, and I was honored to know him.

It is interesting that in those days, well more than fifty years ago, the discipline of biochemistry was largely nutritional chemistry, In Land Grant institutions such as the University of Arizona, the department was most often situated in the College of Agriculture. It was because of this that I got to know Art Kemmerer and the other faculty in his department. While I was waiting for my formal teaching assignment, applying for a grant, and writing the two aforementioned manuscripts, I was planning the experiments that I had envisioned for the hypophysectomized tadpoles that I was raising. These experiments were to involve the injection of MSH into tadpoles, just as I had done for my dissertation. Then, after their guanophores had contracted to the point that they were hardly visible, their skins were to be assayed for their guanine content. Presumably, the pigmentary contents of the guanophores were purines, notably guanine. I knew that I would need help in the extraction, separation, and measurement of guanine, so I turned to my friends in biochemistry. They graciously offered me one of their graduate students, Saul Neidleman, as a collaborator.

Saul was a bright young guy from New York and not much younger than I. He was broadly interested in biology as well as chemistry. He took to the work I proposed with zeal. I presented him with skins taken from normal, hypophysectomized and MSH-injected hypophysectomized tadpoles. Extracts of the skins were studied by means of ascending paper chromatography, a very important analytical tool in those days. In order to localize guanine on the dried chromatograms, we utilized a diazotation reaction that produced a bright orange spot where guanine was located. It was revealed that, indeed, much less guanine was found in skins from intact larvae than was present in hypophysectomized tadpoles. Moreover, skins of hypophysectomized MSH-injected larvae were practically devoid of guanine. Even more fascinating observations were made on some of the chromatograms that we subjected to UV illumination to first locate guanine spots. To our amazement, in addition to the black, absorbing spots that indicated the presence of guanine, we discovered numerous fluorescent spots whose degree of fluorescence corresponded inversely to the amount of guanine; thus, the greater the amount of fluorescing substances, the lower the amount of guanine, and vice versa. A look at the literature told me that these fluorescent compounds were pteridines, and thus I learned that their deposition in the skin is MSH dependent. This discovery opened a major area of my subsequent research endeavors that I think of as "the pteridine period."

During these early weeks and months in Tucson, we were distracted—Lou with her new job and I with the start of my research—from a melancholy that we brought with us. Tucson was not our first choice, for we would have much preferred to be in Baltimore. Moreover, we had each been raised in country that was green and forested. The arid southwest and the Sonoran desert were different and strange. With time, we each began to appreciate our new environment, in large measure because our youthful enthusiasm led us to explore our new surroundings. Not being encumbered by a family gave us more time for exploration. In my case, a great source of enthusiasm came from an association with the principals of the Wildlife Unit, Lyle Sowls and Roger Hungerford. This unit was a federal entity associated with the Department of Zoology. Its head was Lyle and Roger was his associate. Roger, who was just finishing his PhD, taught the undergraduate courses in wildlife management, while Lyle administered the various research programs that the unit was involved in. When the Zoology Department moved into our new building in the spring of 1957, the office I was assigned to was adjacent to that of Roger Hungerford, and we were both near the wildlife teaching laboratory. Lou and I became close friends of Roger and his wife, Carol, and their two children. Our two families partook of many outdoor activities together, from picnics to camping. Similarly, Lou and I developed a strong friendship with Lyle, his wife Grace, and their two children.

As is usually the case with wildlife people, Roger and Lyle were enthusiastic hunters, and I was invited frequently to accompany them to hunt for quail or doves. I accepted because I loved to be out in nature, but I really couldn't stomach the act of killing these birds, even though at the end of the day we would eat them. Gradually, I stopped carrying a gun when we went out together, and eventually I gave my shotgun away. I took many other trips with Roger and Lyle that were associated with their respective research programs. One trip I remember with amusement. One of Lyle's projects concerned a study of wild turkeys. The study was being made by one of the unit's graduate students who was spending the summer high in the Graham (Pinaleño) Mountains in southeastern Arizona. Lyle, Roger, and I left early one morning in June to drive the Swift Trail to the top of Mt. Graham. When we got a good way up, an enormous black bear crossed the road just ahead of us. Immediately, we pulled the pickup truck to the side of the road and took off on foot to chase the bear. Of course, it soon outdistanced us and disappeared, at which point we looked at one another with laughter as the same thought occurred to each of us: What would we have done if the bear had turned around and started chasing us!? I am grateful to Lyle and Roger for having taught me much about the plants, animals, and ecology of Arizona.

Toward the end of the second semester, the Departments of Microbiology and Zoology began our move to the new building, Biological Science, now known as Biosciences East. The move was piecemeal since we all had courses to teach. I was in the middle of teaching my very own course independently. It was an exciting challenge, but not a daunting task since I patterned my course after the two embryology courses that I knew well, that of Hans Holtfreter at Rochester and Emil Witschi's course at Iowa. Each had been taught from a different perspective, and I distilled the essence of each to develop my own version. I was proud of my course, and I feel that it improved with age over the next forty years or so. During the years, I kept up with the literature and found new material to add at almost every successive offering of the course. When the semester was over, I organized my research lab in the new building and moved my tadpoles and whatever equipment and supplies I could from the old department. Since my appointment was for the nine-month academic year, and I had no plans for summer teaching, I had a fair amount of free time until September when I again began teaching. The teaching schedule was adjusted so that I could teach vertebrate embryology in the fall and experimental embryology in the spring to coincide with the amphibian breeding season.

4

The Stadium Years

During that first year at the university, I did not forsake my proclivity for exercise, and I was at the gym frequently, often playing basketball with other faculty and staff members. I believe that it was through such contacts that I learned there were openings for two head residents in the men's dormitories for the forthcoming year. Since Lou and I were not happy with our living accommodations, the head resident possibility seemed attractive. We applied for and were offered the head residency of East Stadium Dormitory, a small dormitory on the first two floors of the northeast wing of the football stadium. In those days, the stadium was a simple horseshoe, and the east side of the stadium bordered on Cherry Avenue, which was a through street at that time. The head resident's apartment was at the very end of the north end, and the private entrance opened onto a concrete apron that served as a stadium entrance for games and as parking at other times. The advantages were great. First of all, the dormitory had few residents and most were serious engineering students who were much less prone to mischief than most college students. This eased the responsibility load. The financial advantage was considerable since we had free rent and free utilities. Access to the campus and all the university activities was just a simple walk across the athletic practice fields. I did not need a car, and Lou reached Infilco easily since it was located on South Campbell, easily accessed from Cherry Avenue.

The apartment was comfortable. It had a large living room with windows facing Cherry Avenue and a small but effective kitchen. The bedroom was actually under the stands and thus had no windows, making for sound sleeping. Noise level was minimal since the apartment was separated from the student quarters by a large leisure room associated with the front and only entrance to the dormitory. We were comfortable, and we lived in the apartment from September 1957 to the spring of 1961. While many would think this to be a bizarre living arrangement, and I suppose that it was, we took advantage of its access to the stadium. The football field gave us an enormous lawn that was more or less private. Most of our friends in those days were young and had children. Of course, they loved to come to visit and cavort on the grass. I had brought a football with us when we came to Arizona, and I enjoyed tossing it around with some of my dorm residents, all males in those days. Footballs in the 1950s were still rather fat, unlike the thinner streamlined footballs of today that are designed more for passing. The fatter football was more conducive to the kicking game, and it was only a few years earlier that the drop kick had become illegal for kicking field goals and extra points. I had grown up with the drop kick, so I enjoyed going out to the "back yard" to kick a few through the uprights just for a little relaxation. These memories are a little bitter-sweet, for now the stadium surface is no longer grass. It has been replaced by an artificial turf as part of the continuing evolution of the now modern facility. Early on, Cherry Avenue was closed and became engulfed by a new bank of stands, and later the remaining stadium structure was absorbed into new stands.

Home football weekends were quite an event. Since we went to the games, almost all at night, we were not disturbed by the noise. Fortunately, the Sunday morning cleanup didn't start until we were up, so Manny and his gang didn't bother us. In fact, it was rather amusing to watch as they collected the empty liquor bottles that the fans had left. As Manny's sweepers cleaned row after row, they placed the empty bottles on the seats and afterwards an official count was made. I have no idea of whatever purpose the liquor bottle census served. It was not illegal to take alcoholic beverages into the stadium in those days. In the spring the stadium provided a venue for track and field competition, and this was to my liking for I was, and am, still a track devotee. In fact, I was friendly with Carl Cooper, the track coach, and he often asked me to officiate at track meets. Every afternoon, members of the track team practiced in the stadium, and I became acquainted with a few of the runners. One of the distance runners stood out, not because he outdistanced the others. On the contrary, during his first year or two he always lagged far behind at the competitive meets. This young man was George Young who, by his senior year, had become a fine runner at the international level. He represented the United States in the steeplechase at the 1960 Olympic Games. These games were held in Rome that year, and Lou and I had the good fortune of seeing George participate there. George had a long, successful career and participated in several subsequent Olympic Games.

The fact that we were in Rome in 1960 was a highlight of our young lives. There was an important meeting on developmental biology to be held in September at Pallanza on Lago Maggiore in northern Italy, and the NSF was offering travel awards to participate in it. I decided to apply for an award, and by just dumb luck, my request was approved. We were thrilled at the prospect of visiting Europe and made plans for an extended trip. In mid-July, there was another international meeting of importance that was scheduled to take place in Copenhagen. It was a congress of endocrinology and included presentations by important researchers whose work was directly related to my own. We gave serious thought to the possibility of participating in both conferences and to spending the two-month period between the two as tourists in Europe. We had no idea what the future would bring or if we would ever again have the opportunity to visit Europe. In those days, Europe was a bargain—hotels were cheap and meals not too expensive if one were careful. Money could also be saved by purchasing a European car for local delivery. We found that a new Volkswagen Beetle could be purchased for $1100 at the factory and for $1200 in one of the larger cities where we would be. When we learned that Infilco would give Lou a leave of absence for the ten weeks that we would be away, it was not a difficult decision for both of us to go rather than my going alone for just the week or so duration of the developmental biology congress. Our decision was made easier because we could make the trip without going into debt. We each grew up in families where savings were emphasized, thus by saving Lou's Infilco salary and by not having paid rent and utilities for the previous two years, we were in good financial shape. This was pretty amazing, because in the spring of 1960, we also purchased the three-acre lot on which we would build our home a year later.

The four plus years of our residence in East Stadium Dormitory were an important time in my development as a research scientist. On the one hand, I continued to follow the lead of my dissertation research on the role of MSH in the regulation of the bright-colored pigment cells, especially guanophores, and at the same time I continued where the tail-darkening reaction of *Xenopus* took me. The latter required that I establish a breeding colony of *Xenopus* of my own. Thus, soon after we were comfortably situated in the new Biological Sciences building, I had my friends in the Witschi lab in Iowa City send me some of their adults. I was in business.

As a consequence of our discovery that MSH affected not only guanophores and guanine content of the skin, but pteridines as well, the question arose as to where the latter were localized in the skin. We knew that the dorsal integument of the leopard frog and its tadpoles contained not only guanophores, but yellow pigment cells as well. Often the latter were superimposed upon the former, prompting some early zoologists to refer to the two as one pigment cell, the xantholeucophore. My observations led me to conclude that we were indeed dealing with two separate cells, the guanophore and the yellow pigment cell, which I called a xanthophore. Some years later I developed a pigment cell (chromatophore) nomenclature, which seems to have withstood the test of time. In any case, when I examined the pigmentation of one of our local frogs, the canyon tree frog, *Hyla arenicolor*, I soon found strong evidence that the pteridines were located in xanthophores. The inner surface of the legs of this frog possesses very thin skin populated exclusively by xanthophores. Chromatographic analysis of extracts of this skin revealed an array of pteridines and led to the conclusion that these compounds were the true pigments of xanthophores.

At this time, I turned my attention to the possibility that the biosynthesis of pteridines and purines such as guanine were interrelated and possibly mediated by MSH. Such a possibility was suggested from my earlier experiments with Saul Neidleman and supported by the molecular resemblance between pteridines and purines. Both purines and pteridines contain two rings, one of which is a pyrimidine. In purines, the second ring is a five-membered imidazole group, and in pteridines this is replaced by the six-membered pyrazine ring. As I began to focus on the problem, I first sharpened my skills in chromatography and became more acquainted with the excellent work being done in integumental pteridines of fishes and amphibians by Tadao Hama and his laboratory at Keio University in

Yokohama. A bit later, Lee Stackhouse, my first PhD student, joined the lab, and he worked on the purine/pteridine problem for his dissertation research. Notwithstanding his laudable efforts, the role of MSH in affecting the mutual biosynthesis of these chromatophore pigments remains unresolved. Perhaps I was guilty of not pursuing this problem further, but as is so often the case, I was distracted by the lure of more expedient and less esoteric research targets. The problem remains unfinished business.

At the end of the summer of 1958, Wayne Ferris, a new addition to our faculty, arrived from the University of Chicago. I was very much pleased, for I had fervently urged that he be hired to fill our need for an electron microscopist. I was aware of some of his publications with Paul Weiss that had dealt with the ultrastructure of the skin. Moreover, I knew that he had a laboratory background not unlike my own. We began a friendship and collaborative interests that persist to this day. I owe Sam, as he has been called since his Chicago days, an enormous debt of gratitude. His presence and help have contributed greatly to my success in research. This will become evident as this narrative proceeds.

When Sam and his wife Edel arrived in Tucson, we became friendly early on, and they soon became aware of our weird apartment in the stadium. It was for this reason that Sam called early Sunday morning, November 16, 1958, with the startling words, "Look out the window." He knew that our bedroom under the stands had no windows. To our astonishment, when we looked out, we were confronted by a mantle of white. During the night, Tucson had had an eight-inch snowfall that covered everything. The football field was covered so deeply that the pole vault and high jump pits were completely hidden, leaving only the goal posts exposed. The palm trees looked strangest of all for their lower fronds hung down weighted by wet snow, while the stiff newly emerging fronds stuck up straight. Our two cats, "Momma Kitty" and "Harry," were really confounded to have their play field covered and with no access to the jumping pits. Of course, we soon had our camera out to record the university campus blanketed in snow.

While the grand snowfall was in part a joyous event, it was at the same time a tragedy. The day before, November 15th, a small troop of Boy Scouts left for a hike in the Santa Rita Mountains. The main trail elevates toward Josephine Saddle, a feature that separates the two major peaks, Mt. Wrightson (Baldy) and Mt. Hopkins. Rain turned to snow as the scouts were ascending, so part of the group turned back, but the remaining few persisted upward. As darkness descended, this group lost their way in the area of Josephine Saddle and descended into a side canyon on the south side of the saddle where they perished from exposure. It was a sad day for Tucson. Over many years afterwards, we hiked to Josephine Saddle often and were always touched to see the modest marker and flowers attached to a tree in memory of the sad event.

A first demonstration of a role for the pineal

When I published my paper on the tail-darkening response of *Xenopus* tadpoles in 1957, I had no idea that this publication "scooped" a similar endeavor being pursued at the University of Utrecht in the Netherlands. Soon after the publication, I had correspondence with Tom Burgers, one of the Dutch investigators who were working on the project. We carried on a friendly correspondence, and I learned that Tom was planning to come to the USA to do some research at the Hormone Research Laboratory at the University of California in San Francisco. In order to see a little of America, he decided to take a Greyhound bus across the country and to make stops at a few labs along the way. His first stop was in New Haven to visit Aaron Lerner at Yale University. At that time, Aaron's laboratory was riding a high after having elucidated the structure of alpha-MSH. This was a significant achievement, especially since it took place in a department of dermatology. Their prowess in chemistry again revealed itself just before Tom Burgers' visit. I present the story briefly.

Like many curious dermatologists, Aaron was in pursuit of knowledge about the disfiguring skin ailment, vitiligo. He reasoned that because the lesions of this disease are often bilaterally symmetrical, the nervous system might be implicated. In an examination of the older literature, he found a paper by C. P. McCord and F. P. Allen published in 1917 that reported the melanophore-contracting activity of beef pineal glands that were fed to leopard frog tadpoles. They dismissed the activity as an unexplained pharmacological reaction and let it go at that. Some thirty-five years later, the results were confirmed by others. When Aaron learned of the work in the 1950s, he reasoned that since the pineal is of neural origin, perhaps it contains neural factors that might have some dermatological bearing. Thus, the Lerner team began to purify pineal extracts in search of components that might cause melanophore contraction as an explanation for vitiligo. In 1958, they purified and identified such an active substance, an indole, that they named melatonin. When Tom Burgers arrived in New Haven, Aaron kindly gave him a small sample of this melatonin which he brought with him when he visited us in Tucson. We were his last stop before going on to California, so he was willing to share a little of the melatonin sample with me. We were both interested in testing the melatonin, so I drove out to Sabino Canyon to collect some leopard frog tadpoles for the test. In those days, leopard frogs were present in numbers in lower Sabino Canyon; moreover, the road was open and free of major restrictions. When we got back to the lab, we found that indeed, when we sprinkled a practically negligible amount of melatonin into a small bowl of tadpoles, they began to blanch within a few minutes. Moreover, just as McCord and Allen had observed so many years earlier, the blanching was due to profound contraction of dermal melanophores while the less prominent epidermal melanophores were unaffected. I don't remember if Tom gave me the rest of his melatonin sample or if we shared what he had. In any case, I did have this small amount of melatonin in my possession for several years before it became famous and ultimately became available commercially. Tom stayed with us for a few days before moving on, and we found the visit memorable. As we sat on the sofa in our little East Stadium apartment, we were mesmerized as we listened to Tom tell us about his visits to some tribes in Africa. Our impressions were enhanced even more as he showed us his collection of photos depicting puberty and circumcision rites of the juvenile tribesmen.

In the late 1950s and for years afterwards, an amazing amount of funding was available for a variety of teaching programs. In the summer of 1959, I was asked to participate in a high school teaching enhancement institute. My task was to present some simple laboratory experiments that the teachers could take back to their respective biology classes. I demonstrated several possibilities, including the *Xenopus* tail-darkening experiment. Having done so was a fortuitous move on my part and was just another example of the place of serendipity in scientific discovery. First, I told them about *Xenopus* and the joys of using it as a research animal. In addition, I offered to provide them with tadpoles or adults should they ever wish to do the experiment I was about to show them. After explaining the theory involved, we actually did the experiment in the lecture auditorium using *Xenopus* tadpoles I had brought up from my lab in two glass beakers. After everyone had a good look at the swimming larvae, I put one beaker in a closed drawer in the counter at the front of the room. The other beaker was left on the counter top in full illumination from the auditorium lights. I explained that within about thirty minutes the tails of tadpoles left in the darkness of the closed drawer would be black, while those left under illumination would remain unchanged. Then, off we all went to the student union for coffee while we waited the circumscribed thirty minutes. To our surprise, when we returned, we found that the auditorium lights had been turned off. (Probably it was I who had done it, knowing my own absent-mindedness and compulsive nature.) At any rate, the tails of both groups were black because all of the tadpoles were in darkness for thirty minutes. It all worked out because the class had seen that the tails had been clear at the start of the demonstration. Moreover, just as I had explained would happen, all the tails began to blanch after the lights were turned back on and were completely blanched after the promised five minutes. The teaching experience was a success.

In doing this demonstration, I made an observation that had escaped me in the many times I had watched the tail-darkening response on whole, intact larvae. When we returned to the auditorium, I noticed that while all the tadpoles had darkened tails, those in the beaker from the closed drawer were slightly less dark than those in the desktop beaker. Moreover, the body melanophores of the tadpoles from the drawer were somewhat contracted, more so than those of the desktop tadpoles. I didn't think that any of the class participants noticed these differences so I said nothing about them, but my own mind was at work. Over the successive months I developed an explanation and carried out a series of experiments to confirm the important hypothesis that I had developed.

During the months that I was concerned about these changes of relative expansion or contraction in light or in darkness, considerable interest in the pineal gland had evolved. In part, I suppose that this was due to the bloom of electronmicroscopy. When the structure of the eye was examined at the ultrastructural level, the lamellar arrays of the rods and cones were discovered and were implicated as the repository for the visual pigments. Similar arrays of lamellar structures were found in that component of the lower vertebrate pineal that became known as the pineal eye or the third eye. This was quite appropriate terminology for in one of the best examples, the pineal eye of the lizard *Sphenodon,*consists of a cornea and lens, and a retina which is connected to the pineal body by a nerve much as the right and left optic nerves connect to the optic tectum. As an embryologist, I followed all of this with interest, because I knew that the pineal gland originated from two half-primordia located respectively on the neural folds, just lateral and in contact with the left and right optic vesicle primordia. Subsequently, when the neural plate folds to form the neural tube, the two half-primordia unite at the midline producing the epiphysis cerebri or the pineal body. Thus, the pineal and the lateral eyes have a common embryonic origin and both function as photoreceptors.

As I began to look for an explanation of why the body melanophores of the *Xenopus* larvae contracted during the demonstration for high school teachers, I had a "eureka moment." Obviously the tadpole blanching involved photoreception; however, unlike the tail darkening in which tail melanophores expand slowly when placed in darkness and return to the contracted state within a few minutes, the time factors for the body-blanching reaction were just the reverse. Body-blanching (melanophore contraction) began within a few minutes after the tadpoles were placed in darkness, and their return to the fully expanded state after the lights were turned back on required about an hour. The expansion and contraction of tail melanophores in darkness and light were clearly based on production and destruction of a visual pigment system, but such a phenomenon could not account for the onset and termination of the body-blanching response. Time factors for the latter in fact suggest perception of darkness by a photoreception system followed by the quick release of a stored melanophore contracting substance. Then, when illumination was restored, the substance would no longer be released and the amount already present in the circulation would be gradually metabolized, leading the melanophores to return slowly to the expanded state. Since I knew that the pineal was a photoreceptor, and that it was also a repository for melatonin, a potent melanophore contracting agent, an hypothesis suggested itself. In the intact larva, melanophores are normally expanded due to the action of MSH, but when the tadpole is in the dark, the pineal is stimulated to release melatonin into the circulation at a level sufficient to override the action of MSH, resulting in blanching. This was a bold hypothesis that needed testing, so I went to work. During the fall and winter of 1959–1960, I did an array of experiments involving pineal extirpation and the use of the small amount of melatonin that I had left from the Tom Burgers visit. All the results led to the substantiation of my hypothesis. Thus, in the spring of 1960, I submitted a manuscript to *Science* entitled "Pineal regulation of the body lightening reaction of amphibian larvae." Of course I was filled with pride when it was accepted for publication, not only because of the prestige of *Science* magazine, but also because the research was the first demonstration of a role for the pineal in a normal, everyday physiological event. As icing on the cake, *Science* also accepted a photo I submitted for the cover of the issue that contained the article. It was to be the November 18, 1960, issue.

5
Going to Europe

Perhaps I am overly exuberant in describing my discovery about the function of the pineal. After all, the body-lightening phenomenon is found in amphibian larvae and in fishes and, in the total scheme of things, this knowledge seems pretty esoteric. However, given the long history of mystery that has shrouded the pineal, including speculation about its position as "the seat of the soul," to discover a legitimate and definitive function is important. Its presence in the most primitive fishes and its evolution up the vertebrate scale has long attracted the interest of scientists. Thus, the demonstration that the pineal serves as both a photoreceptor and an endocrine organ was greeted with an enthusiasm that generated new research about circadian rhythms. The daily cycles of serotonin and melatonin production and their release are now well known, but not many investigators are aware that the roots of this knowledge stem from experiments done on simple amphibian tadpoles. To some extent this may be my fault for not continuing work on the pineal. Instead, I became distracted by other interests and investigation of the pineal became another item of unfinished business that dots my career in biological investigation.

As summer approached, our attention became focused on our trip to Europe. We had done some shopping for clothes and were going to be well dressed for our long visit. Traveling in those days was not the casual affair that it is nowadays. We were well dressed on the plane, and neither we nor others wore jeans. We both looked forward to our great adventure, but my spirits were especially buoyed by the knowledge that while I would interact with scientists in the countries we would visit, my pineal manuscript was in press. Departure for me would be easy since I was unencumbered, but Lou worked until we left. In fact, I recall that she worked at least a half-day on the date of departure. In those days, there were no jet liners. Our overnight flight to Chicago was prop-driven, and the TWA Constellation that we flew to Amsterdam, via Shannon for refueling, was also prop-driven and slow. The trip was further lengthened by a five-hour layover in Chicago between flights and by a change of plane in Amsterdam where we boarded a KLM flight to Copenhagen. By the time we reached Copenhagen, we had been *en route* for some thirty-six hours. We arrived in the early evening, dead tired. Now came another challenge. When we planned the trip in Tucson, we found that we could not make hotel accommodations at a price we could afford. Instead, upon arrival we were to proceed to the central railway station where we could find housing through the housing bureau. When we looked at what was available, we chose a room in a private home. Then, off we went in a taxi to Jagersborgade and the home of the Andersens.

Mr. Andersen was a retired gentleman who drove a horse-drawn carriage in the Deer Park. Mrs. Andersen, a robust woman of about seventy years, was a homemaker who had studied English for her own enjoyment and for the purpose of communicating with roomers. Their apartment was very modest, to say the least, and we had the best bedroom. In fact, it was their personal bedroom that they turned over to roomers during the tourist season. Mrs. Andersen was very kind and informed us that she would provide us breakfast at a price that was much less than we would pay in a hotel or restaurant. We agreed of course, for at that point we were so tired that we would agree to anything! The fact that the room with a continental breakfast of coffee and pastry was only twenty-one kroner (about $3.00) was very appealing. We were ready for some much needed sleep, and when we undressed and were ready for bed, we couldn't believe how swollen our feet were after having been up for so many hours. As we went to bed at almost midnight, we were amazed to see that it was still light outside for, while we were not quite in the land of the midnight sun, we were pretty far north. As we looked out at the neighborhood, it seemed a bit dreary and that, together with the lack of amenities offered by the apartment, led us to believe that the next day we would find something better. In the morning, Mrs. Andersen had a wonderful breakfast waiting for us and was altogether charming and interesting. The world looked much better after a good night's rest so when Mrs. Andersen asked if we would stay the week, we agreed. The lack of a private bath was a concern, but we met friends at the Endocrinology Congress who were staying in an excellent hotel, and we accepted their offer of a good shower now and then during our stay. Our second breakfast with Mrs. Andersen provided us with some humor that we have relished for many years. As we were well engaged with our morning coffee, Mrs. Andersen, who was sitting with us, asked if we minded if she smoked. Of course we agreed, and she proceeded to light up a big cigar, a rather fat club-shaped affair. We could hardly contain ourselves, but we only smiled with the realization that this was a Danish habitude—older women smoked cigars. The remaining pleasant breakfasts followed this same pattern including genial conversation after the meal and before we left for the day. On one of the days of our stay, Mrs. Andersen offered us a ride in the Deer Park in Mr. Andersen's carriage. It was a real Danish experience. As the week progressed, she told us that we must sign her guest book before we left Copenhagen. She also informed us that only guests she liked were asked to sign the book.

The congress was a grand success. I met many new people and got to see the faces of famous investigators whom I knew only from the literature or from correspondence. My PhD mentor Emil Witschi was there, and in fact, he and Lou and I had a pleasant Danish lunch at a restaurant on the Langelinie along the waterfront, not far from the famous "Little Mermaid." Among the people we met at the congress was Marrietta Isidorides, a Greek scientist who was just a few years older than we. Marrietta and her husband had been graduate students in Iowa City and, in fact, had left just a few months before I arrived in 1952. By coincidence, she had lived in the same apartment that we rented after we were married. We quickly became friends, and it was she who offered the use of her shower at the elegant Hotel Angleterre where she was staying. Soon the Congress was over, and it was time to leave Copenhagen for Zurich, our next destination.

The morning we left, it was a bit hectic at Mrs. Andersen's as we prepared to take a taxi to the airport. As we were making our farewell embraces, Mrs. Andersen brought out a gift for us, a small dish by Bing and Grondahl, a real Danish souvenir of consequence. In just this short week, we had become friends. Soon after we arrived at the Copenhagen Airport, we were shocked to hear our name paged. We responded with some concern, but were soon relieved when Mrs. Andersen's voice told us that she wanted us to know that she had forgotten to have us sign the guest book. She did not want us to think that we were not dear guests! This was one more reason why we became enchanted with Copenhagen and with the Danes. We remained in annual contact with Mrs. Anderson for many years until we learned from her son that she had passed away.

The next major step on our European adventure began with a SAS flight to Zurich. There we picked up our *mausgrau* Volkswagen Beetle, a wonderful means of transport for the next two months. Our previous cars, all used, had particular names and the same became true for our brand new Volkswagen which we christened "Liz" after the young woman who delivered the car to us. We had another reason to remember Liz in that she gave us a "*fahrtenbuch*," a kind of small diary to be kept in the glove compartment. It served to tell us where we were when we bought gas, how many miles (kilometers) we had gone, the kind of gas we bought, the date, and so on. When the "*buch*" was full, we found substitutes, and since that day we have had *fahrtenbuchs* for all the cars we have owned. These *fahrtenbuchs* have provided dandy records. Our mouse gray colored little car stood out among the many other gray cars because it had a beautiful Swiss license plate with the broad red stripe of a temporary plate.

So off we went through Switzerland and on into Germany. Traveling was easy in those days because there were not many cars or other tourists to contend with. In Switzerland, we soon learned that by finding a good *gasthaus,* we could have a clean, inexpensive accommodation with good local cuisine. We found the same in Germany aided by the fact that I could speak a little German. In those days, much less English was spoken in rural areas than is the case nowadays. A pleasant memory for us was an evening we spent in a quaint little inn in the town of Hausach in the Black Forest. We were the only guests, and the proprietor hovered nearby as we ate our dinner. We exchanged small talk, even with my pigeon German. In our conversation, I realized that he thought we were Swiss. When I corrected him, he uttered words that have amused us ever since, "*Aber der Wagen ist Schweiz*" (But your car is Swiss). He had seen our Swiss license plate.

During the ensuing days, our itinerary was to be generally north through southern Germany. We went directly from Hausach to Alsace where we were to meet a distant cousin of Eldon's, Odette Richert. She and her mother lived in Bouxwiller, a village about twenty kilometers from Strasbourg. Madame Richert, a widow, and Odette owned an *epicerie* in Bouxwiller and lived in an apartment above the store. Odette spoke English rather well, having studied it at school and honing her skills during a prolonged visit to the USA. She had been invited to come to Iowa by an affluent relative to attend a wedding in western Iowa. During her stay, she had visited Dorothy and Eldon in Cedar Rapids and so the connection was made. Mrs. Richert was a lovely woman with snow-white hair, and Odette was an attractive young woman in her early thirtys. During WWII when the Germans invaded Alsace, they conscripted the Richert son, and he was taken off to fight on the Russian front where he perished. Thus, the two women were left to fend for themselves. We had not intended to impose upon Mrs. Richert and Odette; rather, we stopped to greet them on Dorothy and Eldon's part. However, Odette insisted that we stay for dinner and spend the night. She was kind enough to show us that part of Alsace and to take us to Strasbourg for a visit the next day. At that time, Bouxwiller and the nearby villages looked old for they had not fully recovered from the war. Over the fifty plus years since our first visit, we have had the good fortune to return to Bouxwiller and Alsace on numerous occasions. This is a wonderful part of France, and Bouxwiller with its beautiful half-timbered houses (*colombage*) has been fully restored. Like many Alsacian villages, Bouxwiller has an ancient stork nest and resident storks. It is a real jewel among the many picturesque villages of Alsace although it is off the usual tourist beat, probably because it is not on "*le route du vin*" (the wine road).

Over the years, we have gotten to know the *route du vin* and its associated villages rather well. The first visits we did on our own, but later we went with Odette and George Hettinger, a widower she married a bit later in life. When we first met George, he had retired as a *brasseur* (brew master) of a brewery in Ingwiller, a village six kilometers from Bouxwiller. By this time, Madame Richert had retired, and the epicerie had closed. Odette and George were living in the beautiful house George had built near the brewery in Ingwiller. We had the good fortune of being their houseguests on several very pleasant occasions. George was a dear person, kind, sensitive, and possessing a wonderful sense of humor. He knew all the better inns and restaurants in the area and on the *route du vin*, and he was equally well known to the proprietors. What fine meals we had together. When we were with George and Odette, we truly affirmed that the Alsatians are unique, that they had taken the best from Germany and the best from France. Our impression is that they are Alsatian first, but are also proud to be French citizens. As time has gone on, the Alsatian dialect is heard less and less. It seems that only the very old can speak it. George and Odette were both Alsatian speakers, but they spoke French to one another. We spoke little English when we were with them. Odette was less comfortable with English as she got older, so we speak in French. She now writes to us in French, although we write to her in English.

After we left Bouxwiller and progressed north along the Rhine, our next fixed destination was Arnhem in The Netherlands. Here we met Tom Burgers in his hometown where he lived with his mother and father. They were gracious people who we very much enjoyed despite a bit of a language problem. Tom, of course, spoke perfect English, but this was not true of his parents. Nevertheless we muddled through with English and a word of two of Italian that Mr. Burgers had picked up in Italy. He had been a furrier before WWII and had made trips to Italy to purchase furs. As we enjoyed our visit, we learned why they had refused to speak German with us despite their being fluent in the language. During the war their elder son, Tom's brother, had operated a clandestine radio station that he and his friends utilized in order to help fallen British pilots make their way home. The occupying Germans located the station and took the son away to be executed. The parents were further punished when the Germans destroyed the fur shop. It is no wonder that the Burgers family refused to speak German. We found that most Dutch people of that age also refused to speak German.

When we left Arnhem and Holland, we moved across Belgium and on to France and our next fixed destination, Paris. What a thrill it was when we approached the "*bassin parisien*" and looked down at the city from a distance. From our vantage point the famous monuments, the Eiffel Tower, l'Arc de Triomphe, and Sacré Coeur stood out. We had looked forward to the moment not knowing that in ensuing years Paris would become a second home, a city that we know as well as we do Tucson. That first view was much different from what we would have today, for now modern skyscrapers dilute the classic panorama. When the awe began to diminish, we made our way to rue St. Honoré where we had made the first advance room reservation of the trip. We chose this rather pricey hotel, Hotel de France et Choiseul, in an elegant location at the recommendation of Bill and Elena Brown. They had been in Europe the previous year on sabbatical and had enjoyed the hotel. Now, in retrospect, we would never choose this hotel, because it is on the right bank of the Seine, and we are left bank people! Actually, the hotel was quite convenient and close to many of the places we wanted to visit—the Louvre, the Eiffel Tower, the Arc de Triomphe, Place de la Concorde, Notre Dame Cathedral, the Champs-Elysées, and many more famous sites. Of course we liked Paris, but we came to love it in later years.

We left Paris for Normandy and soaked up as much as we could, then continued east, through the Loire Valley, and on into Switzerland. We crossed the Simplon Pass into Italy and went on to Milan for the night. Back then, in 1960, my ability with the Italian language was quite rudimentary. Had it been better, we could have better understood the heated conversation that was taking place on the other side of a hedge in the garden restaurant where we had dinner. What I did gather from our eavesdropping was that this was probably a married man having a lover's quarrel with his mistress. They soon left, and we went on with our dinner. I had asked the waiter for a local wine so he served us Bardolino that we very much enjoyed. As we headed toward Venice the next day, we saw frequent signs advertising something named "Borsalino," and since I had forgotten the name of the previous evening's wine, I assumed that Borsalino was the name of that wine. Because we were traveling by car, we took a hotel in nearby Mestre from which we could then take a boat to Venice. That evening we had dinner in a Mestre restaurant where I proceeded to order a bottle of Borsalino with our dinner. The waiter looked at me with astonishment and then broke out laughing. Obviously taken by my naivete, he kindly explained that "Borsalino" is a brand of hats, and that what we wanted was Bardolino. We had a good laugh to end the day.

We made short work of Venice largely because we had a bad experience with a group of German tourists *en route* back from Venice to Mestre. We had just missed a waterbus leaving Venice so we were first in line for the next departure to Mestre. As we waited patiently, other passengers began to arrive including a fairly large group of Germans. As the waterbus arrived, the Germans pushed forward, ignoring those of us who were waiting. I would not put up with this behavior, so I extended my arms and pushed the swarm aside to allow Lou to board, and I followed. There were a few protests from those rejected, but I was a pretty substantial fellow in those days, so there was no further argument. Over years of travel in Europe, we came to learn that this was common behavior. Nevertheless, such aggressive manners and the failure of Europeans to respect lines is something that I have never learned to tolerate. We compensated for this shortcoming with pleasant visits to Venice in later years. On the 1960s trip, instead of going back to Venice a second day, we visited Stra and Padua as we continued our journey south.

We had no hotel reservations in Rome. As we approached its northern suburbs, we spotted a small facility along the side of the road that turned out to be an official clearing house for housing for the 1960 Olympic Games currently underway in Rome. We found that hotel prices were very high and that third-class without bath was all that we could afford. The man we were talking to, Sr. Sgarallino, suggested a very nice private home with a bath of our own at the price of a third-class hotel. It turned out to be his own home located on the northwest side of Rome, not far from the Olympic stadium and quite near the Vatican. This section, known as Monte Mario, is very pleasant. The Sgarallinos were very nice, and they had two adorable little girls. He was a metallurgical engineer who was temporarily out of work and who took on his housing job as temporary employment. We presume that the rental of a room in their home was also a means to provide income while he was unemployed. Altogether it was a great accommodation during our stay in Rome. Of course we visited as many as possible of the historic sites and monuments, many of which we saw again over the years during subsequent visits.

What was unique about August of 1960 were the outdoor Olympic Games. We were fortunate enough to be able to acquire tickets for two afternoons that became famous for Americans, "black Thursday" and "good Friday." The former because we did not win the events we thought we would and the latter because we won events when not anticipated.

Our drive south to Naples, as well as the earlier segment between Bologna and Rome was much enhanced by a gift Eldon made in anticipation of our visit to Italy. During WWII, he spent a substantial part of his service in Italy. After the North African campaign, he was part of the invasion of Italy and the march north. He was involved in important battles and was again wounded—the first had been in North Africa. During the Italian campaign, he was the map bearer for Company C of the 34th Division of the Fifth Army, and because of this, as well as his own interest, he knew where he was and the names of the geographical sites they passed or dealt with. He prepared for us a small pocket-sized spiral notebook complete with photos of places and maps of areas we would pass or could visit if it fit our itinerary. We enjoyed matching our travels with his when we could. On our drive to Naples, we went by way of Latina to visit a farm where Eldon had been billeted. A woman we spoke to there had been at the farm when the house was occupied by the American soldiers. She refused to look at our photo of Eldon.

Our route to Naples took us along the sea past the beautiful port of Gaeta. We were on the old Roman road, la Domiziana, a road that we would come to know and love in later years. In Naples, we stayed in a hotel recommended by Bill and Elena Brown. It was on the via Partenope, along the bay. It was not elegant, but was comfortable enough. We later learned that it was a hotel used by easy women! What I remember about the hotel was that it had a sidewalk restaurant where we dined one evening, and where we had the good fortune of seeing the U.S. satellite Echo pass overhead. One evening a few weeks earlier, we had gone out into the stadium to watch Echo as it passed over Tucson. The one down side about dining in the open air at that time was that it was the "*festa di piedigrotta.*" This event-dedicated to the Madonna of a cave located just north of Naples-had been a music festival for which many of the famous Neapolitan songs had been composed. Unfortunately the festival had evolved into an occasion where undisciplined youth wander about bopping people in the head with rolled up newspapers.

It was great to see Giovanni and Annamaria Chieffi and their four beautiful children, including Matilde and Lorenzo who we knew in Iowa City. They were now six, five, three, and an infant (and there were two more to come). The visit in Naples was a full one. We spent much time viewing the antiquities, and one afternoon I gave a seminar at the Aquarium (*Stazione Zoologica di Napoli*), a place that I would get to know and love ten years later. One evening was especially interesting and funny. When we arrived in

Naples and found Giovanni, he informed us that he would not be available one evening because that was when Italy and Yugoslavia, the two favorites, were to play their Olympic soccer match. We asked if we could go too, and so we did. It was incredible to witness and hear the Italian fans—the gesticulations and noise were fantastic. The teams played to a scoreless tie, and so they played an extra period. Italy scored first, and this was matched by Yugoslavia as the period expired. At this point the players were so tired they could hardly walk. The winner was then to be determined by what the Italians called "*la moneta*" (toss of the coin). Lou and I looked at one another with the same thought. After all the rowdy behavior that the fans had exhibited all evening long, what will happen if Italy loses the coin toss? Well, Italy lost the coin toss and nothing untoward happened—the fans just got up and left with an attitude of "that's the way it goes."

When our four days in Naples were over, we headed further south. I had promised my father that we would go to his hometown, Delianuova, in Calabria, to meet his two younger sisters and their families. I was really not that enthusiastic about doing so. I didn't know my *Calabrese* aunts and cousins, and my father had been away from Calabria for forty years. Perhaps more importantly I didn't want to witness the poverty that I knew existed there, just as it did in rural areas of many European countries. During my years growing up, I remember well the frequent "care packages" that my parents sent to Delianuova. Nevertheless, I knew that I had to go. My father was retiring the next year, and he wanted to return to Italy to visit his sisters just as soon as he could be free. In 1960, the trip to Delianuova from Naples was a long one because the wonderful Autostrade del Sole had not yet been built. We took it to Reggio Calabria about ten years later and several times afterwards and marveled at the engineering feat it represented. It made what was an overnight trip into one of only a few hours. Our first trip was even longer because we went by way of the beautiful, but slower Amalfi drive, to Salerno. The rest of the drive down to southern Calabria pretty much followed the coast, but was often tortuous when some of the hills came down to the sea.

As we left Salerno going east and then south, we entered a plain formed by the Sele River. It extended to Paestum, one of the more important archeological sites in Italy. Paestum was first inhabited by the Lucanians, the indigenous people of the area whose artifacts have become more appreciated in recent years. The Greeks occupied the area subsequently and erected enormous temples that are in a remarkable state of preservation. They are among the more impressive Greek temples anywhere in *magna grecia* or in Greece. The Romans followed the Greeks and added many typical features, notably exquisite mosaic floors. In large measure, Paestum remained so well preserved because the Romans were forced out because of malaria, and the whole area became hidden by thick vegetation. The Sele River enters the Tyrrhenian Sea just north of Paestum, and this whole area is quite moist. This was the source of malaria. In later years, I came to know the area fairly well because of frequent trips to Paestum, and because the Sele River was the source of some of the fresh water lamprey I worked on in Naples in 1970-1971. Paestum was rediscovered during the reign of Benito Mussolini prior to WWII when he had the whole area drained to get rid of the malaria-bearing mosquitoes. Gradually, after the war, considerable archeological investigation and excavation was begun, and a small museum had just been built prior to our 1960 visit. We have been amazed at the continuing progress that we have observed with each visit over the next thirty or forty years. The progress is well documented in the important newer museum that now exists there.

I cannot leave the Sele valley without commenting about one of the principal products of the region—cheese of all sorts, notably *"mozzarella di bufala"* (mozzarella made from the milk of water buffalos). Battipaglia, a small city at the west end of the valley, is well known for its cheese. We were unaware of this as we drove through in 1960, but when we lived in Naples in 1970-1971, we had buffalo milk mozzarella almost every day, and its source was either Battipaglia or la Domiziana. The latter is a large plain north of Naples through which we passed as we first entered town. Water buffalos were brought to the areas I mentioned by the Romans who imported them from Egypt. Buffalo mozzarella eaten the day it is made is divine! For me it was also once a source of great amusement. A few years back, when I was editor of *Pigment Cell Research*, I received a manuscript for publication by a young investigator from the Agriculture College of the University of Naples. He had observed water buffalos (which are black or dark gray) with light colored lesions very much like human vitiligo. The research he described was a well done piece of work, and the reviewers recommended publication with only slight revisions. When the author returned the corrected manuscript, I wrote a letter of acceptance. I was sure that the author had no idea, despite my Italian name, that I knew about Naples and mozzarella, so I added a postscript to the letter asking if the vitiligo affected the quality of the mozzarella, and if the buffalos came from Domiziana or Battipaglia. Indeed, my letter astonished him, and I'm sure "it blew his mind" to think that this American editor could have such knowledge about mozzarella and its Neapolitan geography. In his response, he invited me to meet him for some mozzarella on my next trip to Naples.

We continued on past Paestum and through a tortuous mountain area and back to the sea, to the Gulf of Policastro. It was now time to find a room for the night. Fortunately, when we entered Italy from Switzerland, we had the good sense to join the ACI, the Italian Automobile Club, which had provided a listing of motels and hotels. It indicated that there was a motel in Cirella, a small town on the coast just into Calabria, and so we aimed for it. The driving was easier now as we quickly exited Campania, the province where Naples is located, and after a quick passage through a narrow strip of Basilicata (Lucania), we entered Calabria and soon found Cirella. The next morning, we moved down the coast. I was driving so I could not appreciate the sights as well as Lou, so I quote from a letter she wrote home to her folks: "On Saturday morning we wound on down the coast through Paola and Amantea, first along the sea, then above it, then back down, then up to a town hanging on a cliff. It's quite a drive! Along the Golfo di S. Eufemia it was flat and there we enjoyed seeing the women dressed in a sort of costume with black or dark green overskirts brought up into a bustle. Under this is worn a bright red skirt with either a white band on the bottom or still another, longer white skirt underneath. The costume didn't seem to be for tourists' benefit (there are hardly any) as the women were doing their regular work—washing clothes in the public fountain, tending the children, working in the fields, etc." A little farther south we passed through the town of Vibo Valentia where we spotted a good-looking motel. Since we were not far from our destination, and since I was determined not to stay in Delianuova, we made a reservation for that night—smart move! It soon started to rain as we headed toward Bagnara Calabra where I knew we could gas up. I also wanted to say that we had been there! At that time, I did not know that my surname was not really Bagnara, but Abagnara, as mentioned earlier. I suspect that the name of the port town Bagnara had also undergone an evolution over the years, for some antique maps of the area that I have seen indicate that the site was at one time named Gugnaio. The surname Bagnara is actually a common name in the northern province of Liguria where Genoa is located. Bagnara Calabra was very picturesque, and this impression was confirmed years later during other visits when we descended down to the port where we were much taken by the brightly painted fishing boats. From Bagnara we drove upwards and upwards through woods of edible chestnuts and olive groves that were beautiful even in pouring rain.

Finally, just after noon, we arrived in Delianuova and easily found my relatives. In Lou's words, "What an experience that was!" There were many to greet us, including my two aunts and their husbands, seven or eight cousins, some with husbands and wives, and eleven or twelve children. All these people were in one room around a large table. When one considers that Lou and I were not exactly small people, and that my aunt Concetta probably weighed about 220 pounds, that was one large mass of humanity. Of course, each of them kissed each of us on both cheeks and that nearly did Lou in. I was not in the best shape either for I had developed a splitting headache, a rare thing for me, probably induced by stress. To make things easier, I suggested to Lou that she try to keep names straight while I translated as best I could. All conversation was in the Calabrese dialect, which I could understand fairly well, but I was not a good speaker. I grew up speaking mostly English at home, and that was also the case for my Rochester cousins and most other first generation kids who wanted to be identified as Americans, not Italians. After initial pleasantries were exchanged, we were ready for a giant pranzo. My father had sent my aunts a substantial sum to pay for all the food. Again I defer to Lou's recounting of the meal: "A dinner was brought out and what a dinner. I was too nervous to eat and Joe had a splitting headache so we picked at our food, but the rest ate with gusto. We had chicken soup with great chunks of chicken (mostly skin) and hard-boiled eggs in it (Joe hates chicken skin and I've found that in Europe it comes complete with feathers); pasta with fresh tomato sauce-each was served in a quantity similar to what I'd prepare for an entire meal for the both of us; the meat from the tomato sauce; fried chicken; a veal cutlet; and braciole (a veal steak rolled around ham and hard boiled eggs, and baked); bread; fruit; cake; wine and beer. I'm sure that they don't eat like that all the time, and they may still be eating the left-overs." By the time dinner was over, the rain had stopped and we took pictures of everyone, drove around the town, visited a cousin with a new baby, and returned to Concetta's for a tarantella. Just after dark, we drove back to Vibo Valentia and our comfortable motel. What a day! Yes, people were poor and dressed in well-worn clothes, but at least they had dwellings with several rooms although the toilet was still so primitive. My relatives seemed nice, but we were sad that some were illiterate. Fortunately, they had a new school so the children were going to have a chance for a more prosperous life.

Given the state of affairs in Delianuova at the time of our visit, I strongly urged my father to forego a visit there. He was adamant, however, about going back to visit his sisters for a six-week stay. The next summer he made the trip, and while I'm sure he was happy to see his sisters, he soon became disillusioned and left after a short time to spend the rest of his Italian stay in a pensione in Rome. After my father died in 1973, we lost all contact with our relatives in Delianuova. Years later, in 2003, Natale Sofi, the son of one of my first cousins, made an effort to find me and my sister. Natale is the son of Santa Zappia who was a young woman at the time of our 1960 visit to Delianuova. During those forty plus years, things had vastly ameliorated in southern Italy. All my cousins had good jobs, and they and their families had achieved a certain degree of prosperity. Natale was about forty years old when we first talked by telephone. He had a very good position in Italian telecommunications and had traveled extensively. He could speak some English, but since I had become fluent in Italian over the years, we spoke in Italian. He was a bachelor at the time, and I guess had enough time to call almost monthly for the next few years. After he married and assumed more family responsibilities, we have spoken less frequently. During the period before he was married, Lou and I went to Delianuova on two occasions when we were in Sorrento. There was a time when we visited Sorrento often over a several-year period.

After leaving Calabria, we hurried north for there was much to see, and Pallanza on Lago Maggiore was a long drive away. We had about a week to get there as the International Institute of Embryology Congress was to start in mid-September. On our way north, we made it a point to stop at locations where Eldon had been with the 34th Division. Sometimes we took photos of the very same places that Eldon had photographed. For example, in Pisa, Eldon had taken a picture of a fourteenth century church, Santa

Maria della Spina, and its adjacent blown-out bridge. We also photographed the church and the new bridge that replaced the destroyed one. We showed Eldon's photo to a guard at S. Maria della Spina. He had fought with the British during the liberation and he exclaimed, "That's how it looked when I came home." In Florence, we stopped at the Geographic Military Institute to obtain some maps that Eldon wanted. The people there were really taken by the little notebook he had made for us, and they were impressed by the photos he had taken. We had similar reactions from other Italians we encountered as we experienced vicariously other events of Eldon's passage with the 34th Division.

We arrived at Lago Maggiore in the sunshine, but during the meeting it began to rain heavily for several days. By the time we were ready to leave Pallanza, the lake had risen to such a degree that the front door of the hotel was flooded. We had to use the back door, probably through the kitchens, so that we could depart. We headed toward the Simplon Pass. Before long the sun emerged. To again quote one of Lou's letters: "We had the most beautiful day imaginable for crossing the Simplon Pass: the air was crystal clear, the sky was bright blue, and what a scene the green, green meadows, rushing waterfalls, and snow-capped peaks made!" As we drove through Switzerland and on into France, there was a touch of fall in the air and leaves were beginning to turn color. We drove into Paris in the late afternoon and turned over our car to Panocean, the shipping agent, and they took us to the airport as we started the long flying trip home. Our wonderful two-month stay was over, ending in a whirlwind of activity that got us home to Tucson. We left Pallanza on Tuesday, September 18, drove to Paris, and arrived in Tucson on Thursday evening, September 20. The fall semester started the next week. Unlike current times when classes begin around the middle of August, in those days they began a full month later.

6

Building a House

There was not much going on in the lab while I was away in Europe except that Lee Stackhouse, my first PhD student, was busy with his own things. I put my two cents in upon my return, but I was more focused upon my major teaching responsibility, the vertebrate embryology course. When that was underway, I turned to writing. There were grant proposals to write and manuscripts to prepare. Much of this writing was concerned with the new field known as comparative endocrinology. The emergence of this field was championed by Aubrey Gorbman of the University of Washington and Howard Bern, University of California (Berkeley). Aubrey was an organizer of one of the first comparative endocrinology meetings held in the late 1950s, and he and Howard began collaboration on a textbook of comparative endocrinology shortly thereafter. "*A Textbook of Comparative Endocrinology*" by Gorbman and Bern was published in 1962 and became the textbook that I used in the endocrinology course that I taught. I met both Howard and Aubrey in the late 1950s, and we became friends despite the fact that I was considerably younger than they. Soon thereafter, Lou and I developed a lasting friendship with their families as well. We loved Genevieve Gorbman, just as we love Estelle Bern. In large measure, my own close association with comparative endocrinologists was due to the fact that they are for the most part warm, outgoing people, quite unlike the developmental biologists with whom I also associated.

One of the manuscripts I worked on after our return from Europe was based upon a presentation I was to make at the 3rd International Symposium on Comparative Endocrinology to be held in Oiso, Japan, in the spring of 1961. Little did I know that so soon after our return from Europe, I would again travel overseas. However, when I learned that AIBS (American Institute for Biological Sciences) was accepting applications to fund travel to the Oiso meeting, I applied. Again, with dumb luck, I was successful.

In the meantime, we were much occupied with plans for the building of our house. It would be our first and only home. Thanks to the money we were able to save by being head residents, we were able to buy a lot and begin homebuilding earlier than we would have otherwise. Early on, we began thinking about where we would like to build a home, and at the same time we became certain that we wanted our home to be designed and built by Tom Gist. Tom was a transplanted New Englander who was beginning to make his mark as a builder of unique homes whose designs were much influenced by Frank Lloyd Wright. About this time, we had the good fortune to find a fabulous three-acre lot in the Catalina foothills that was available at a price we could afford. It was owned by Neta and Clay Hubbard, a couple who were dear friends. Neta worked at Infilco, and Clay was an engineer at Hughes Aircraft. Clay found that there were better professional possibilities if he relocated to a Hughes plant in Costa Mesa, and so they offered their lot in Catalina Foothills Estates #3 to us at the same price that they had paid. We followed Tom Gist's advice and bought the lot in 1960 and engaged him to draw up some plans. We loved what he came up with, but our "depression mentality" came into play when we saw the price, and so we opted for a more modest design. We should not have been so conservative, but we nevertheless ended up with a gem of a small house. Ground was broken for the house in April 1961, but before that we made frequent trips to the lot to be certain of the home site and to ascertain which desert plants we might wish to relocate.

On one of our visits, our neighbor from the adjacent lot to the west came over to greet us. He was C. W. Bond who, with his wife, Clarisse, and their two children, spent winters in Tucson and summers in Arlington, Tennessee, where he owned a Ford dealership and had other financial enterprises. When Mr. Bond established that we were acceptable and that we would be available during the summer, he offered us their home to stay in while they would be away and while our house was being built. That suited us wonderfully, so we moved in at the end of the spring semester.

This was the state of affairs when I left for Japan to participate in the 3rd International Symposium on Comparative Endocrinology. Little did I know that just a year after our first trip abroad I would have the honor of another even longer overseas trip. I was looking forward to the symposium where I would have the opportunity to present my work and again see old friends. In addition, I would be able to go to Keio University to meet Professor Tadao Hama and his associates who were working on pteridines present in the skin of fishes and amphibians. The flight from Los Angeles was a long one and again on a prop-driven aircraft. I met Professor Witschi in Los Angeles, and we flew together. Our refueling stop on Wake Island was eventful. When we were ready to take off, a major problem with the plane was discovered. We were all deplaned and taken to some vacant army barracks to sleep for a few hours while we awaited a replacement aircraft to arrive from Japan. We all went to bed after a complicated process to ensure that all sleeping assignments were proper and that pairs of opposite sexes, unknown to one another, were not placed together. No sooner had we gone to sleep than we were awakened by the news that the aircraft problem was miraculously repaired and that we were all to reboard the plane. We arrived at Haneda Airport in Tokyo in the wee hours of the morning, considerably late. Nevertheless, there was a large contingent of Japanese colleagues, all sporting welcoming signs there to meet us. They quickly transported us to Oiso, just west of Tokyo on Sagami Bay, and to our hotel. It was long ago, but I remember every detail of this first visit to Japan in 1961.

This international conference was an important event for the Japanese and was one of the first such scientific conferences to be held in Japan after WWII. Aside from Professor Witschi, I was happy to see many friends among the participants. These included Giovanni Chieffi and Professor Louis Gallien (Big Lou) whom I had not seen since his stay in Iowa City when I was a graduate student. Also present were Howard Bern, Aubrey Gorbman, and Irv Geschwind, all pillars of the comparative endocrinology community and who had become good friends. Among the old Iowa group were Don Johnson, a fellow former student of Emil Witschi, and Shelly Segal, who not only was there for the conference, but to celebrate his honeymoon.

After the conference, I took a hotel in Tokyo for a few extra days in order to tick off a few things I wished to do. One objective was to greet

Ichiro Yoshihara, an executive with Ibara, an important industrial firm that had a connection with Infilco in Tucson. Ichiro had spent six weeks in Tucson, and Lou and I got to know him quite well. I called Ichiro my first morning in Tokyo and agreed to meet him for lunch. I was much impressed when Ichiro arrived at my hotel, The Takanawa Prince, with his private car and driver. I soon learned that in those days executives were provided with many amenities instead of high salaries and so were relieved of excessive income tax. Thus, Ichiro not only had a private car and driver, but enjoyed privileges at the finest restaurants. Lunch was at an elegant tempura restaurant where we had our own private chef prepare the food in front of us as we sat and ate. Ichiro insisted that he pick me up again for dinner. Of course, I did not resist the invitation, and so he and his driver arrived again in the evening. We were delivered to a French restaurant that I am sure rivaled the Tour d'Argent in Paris. Ichiro was an exceedingly gracious host and a stimulating conversationalist. At some point during the course of the evening, he presented me with a train ticket to Nikko that he had purchased during the afternoon. At lunch he had inquired about my schedule in Tokyo. Since I had no fixed plans for the next day, he decided that I should visit Nikko, one of the more important tourist destinations in Japan. He explained that I needed to be at Tokyo Central Station early the next morning to board the train and that the seat was a luxury seat (of course). The only glitch in the plan was that Ichiro did not know that there were only Japanese speaking tours available at Nikko and that English-speaking tours left from Tokyo. I learned this when I left the train at Nikko Station.

Nikko is in Nikko National Park, one of about twenty-five national parks in Japan. It is located in the mountains some 100 miles due north of Tokyo and is situated in a beautiful forest of criptomeria and Japanese maples, a variety of other trees, and thick bamboo groves. Scattered about are many beautiful temples and shrines, including the famous Toshuga Shrine. From the town of Nikko, a beautiful winding mountain road leads to lovely Lake Chuzenji whose outlet is the spectacular Kegon Falls, a long thin waterfall so picturesque that it is often a backdrop for wedding pictures. I have been to Nikko several times since the first visit in 1961, and Lou has been with me twice. During one of my early visits that took place in November, the maples were fully dressed in their magnificent bright red, delicate leaves, and I took a photo that we both love. On one of our last trips to Japan in November 1999, I was able to show Lou the exact spot in Nikko where I took the photo, and it seemed to have not changed one iota.

As impressive as Nikko is in its own right, the 1961 visit was more striking to me as to how it came about. When I descended the train and began to look for a ticket booth, I was informed that I could buy a ticket only for a Japanese speaking tour. There was no recourse other than to buy into a Japanese tour, and so I became a member of the tour "the cat" (after a famous black and white cat carved into the lintel over the door of one of the temples). I bought my ticket and received an identifying cat pin that I attached to a lapel. Our leader was a petite (what else) young lady who sported a typical Japanese hat and who carried a flag with our cat design on it. She held up the flag, and away we went to the bus that was to transport us through the park. Since I was to be language challenged, I decided to place myself near a young Japanese couple, thinking that young people would be more likely to speak a little English. The bus had regular seats with an aisle, and when these seats were filled, an additional seat was folded down from the side, one by one, to make an additional place at the expense of eliminating the aisle. I never saw anything like it; the passenger section was a solid sea of seats! When all were aboard, we drove a short way to some sort of restaurant or teahouse. We descended and sat at tables to drink some tea. There was a lot of conversation between the servers and my tour mates who were being served. I had a feeling that I was missing something, but I just went along with the flow. We soon boarded the bus and were transported to the first tour point. For the next hour or so, we walked to one temple after another. Of course, I could not understand a word that was said, but there was certainly a load of visual input. The time came when we again boarded the bus and were taken to the same facility where we had earlier stopped for tea. I assumed that we were to have lunch, so I made sure that I sat near the young couple. Soon, dishes of food were brought to each of the tour members except me. At this point, one of the young people said to me in plain English, "Aren't you going to eat?" Then it dawned upon them and me that I had not placed my order when everyone else had during the earlier stop. They kindly called over a server and helped me order a simple dish, a curry of some sort. I was served immediately and was offered a spoon to eat it with. At this point, I decided that I was going to be as Japanese as I could, so I conveyed with hand signs that I wished to eat with hashi (chop sticks). By this time, everyone had finished lunch and watched me with great amusement as I downed my curry dish using hashi with a fair level of dexterity. I had been in Japan for almost a week and had developed an ability to eat with hashi. An elderly man sitting across the table from me suddenly started speaking to me in good English. He had lived in Seattle for forty years and had returned to Japan to retire. As we walked from one temple to another, the old man stayed near. Some of the temples required an entry fee. When this was the case, the old man would turn to me and say, "Ten yen!" Among the more famous temples was one that had a beautifully carved, painted lintel depicting three monkeys representing, "see no evil," "hear no evil," and "speak no evil."

When we finished with the temples, we boarded our bus and rode up to Lake Chuzenji and the top of Kegon Falls. Hundreds of photos were taken as all aboard the bus took pictures of one another. The young couple was by themselves so I asked if they would like me to take their picture with the falls as a backdrop. It turned out that they were on their honeymoon and were happy to have me photograph them. Soon we all returned to the Nikko station where the "tour of the cat" dispersed. I had some time before my train was to depart for Tokyo so I accepted the invitation of my new young friends to have a drink together at a nearby bar. It seems that they were from Osaka where the young man was employed as a ship chandler. He gave me his card which I kept for years.

During my remaining time in Tokyo, I visited the Ginza area to shop. Even in those days when we had little money, I felt it important to take advantage of the wonderful buys to be had. Japanese pearls, silk, and optics were the best bargains, and so I bought Lou a length of lovely silk fabric and a beautiful strand of Mikimoto pearls. I needed a decent pair of binoculars, so I bought a pair of Nikon binoculars that I still use after all these years. One of the last days was set aside for university visits and, in particular, Keio University in Hiyashi/Yokohama where I visited Professor Tadao Hama and his associates. In addition to Hama, I spent time with two graduate associates of his, Masataka Obika and Jiro Matsumoto. They were both involved in their respective dissertation research programs. From them I learned much about chromatographic techniques to employ in the separation and identification of pteridines. I left invitations to Professor Hama to visit us in Tucson and to the young men to come for a prolonged stay in my lab. Altogether, this 1961 visit in Japan was wonderful. Little did I know that it was the onset of a life-long association with Japan and Japanese investigators.

Upon my return from Japan, we got our first taste of desert living as we occupied the Bond house until the completion of our new home. It was great to watch our house being built and really fascinating to become acquainted with the beasts and plants that we shared the desert with. They have been a source of my attention and interest over these fifty plus years. To see the changes that have occurred is both a disappointment and a pleasure, as the numbers of old species wane and new ones appear.

As the house project progressed, I returned to the lab. Lee Stackhouse was involved with the pteridine work, and I cleaned up some other projects. One of these concerned the fact that the paired thymus glands of hypophysectomized tadpoles were much larger than those of intact controls. It turned out that these effects were indirect ones based upon the differences in circulating corticosteroids between normal and hypophysectomized tadpoles. This was to be the thesis project of one of my graduate students who began to follow up my observations. He started the project, but when he was accepted to medical school, he disappeared from the scene taking with him the preliminary data that had been collected. Again, my priorities were elsewhere and so the project became another entry into the list of unfinished business.

There was still some work that I continued on the tail-darkening reaction and on the pineal, and at the same time I began to study the bright-colored pteridine-bearing cells more closely. When I was at Keio University, I learned that Masataka Obika had concluded in his doctoral work that pteridines were the definitive pigments in the yellow pigment cells (xanthophores) of the Japanese newt. By entirely different means, I had determined, as pointed out earlier, that these same pigments were responsible for the yellow pigmentation of frogs and an array of other amphibians. From that point on, the bright-colored pigment cells (xanthophores and guanophores) were fundamental targets of investigation for most of my remaining years of research.

The year 1961 was an eventful one. Our house was finished in September. We moved our meager belongings and purchased additional furniture. In addition to all of the indoor projects that needed to done, there was much work to do out-of-doors. The area around the house was relatively undamaged by the building project, but nevertheless needed much work. I was up to the task and particularly enjoyed hard labor. For many years much of my spare time was spent outside.

Our lot is three acres, and the eastern and southern boundaries border on a huge wash. In effect, our actual three acres seem more like fifty. It is perfect for someone interested in natural history, and I spent a fair amount of time there. In addition for providing a venue for bird-watching and hiking, the wash was a source of material for my garden, especially top-soil and decorative rocks. On the far side of the wash to the east is a cliff which years ago had several rather deep openings. One year we were thrilled that a great-horned owl nested in one of the little caves, and years later enjoyed watching a bobcat rear three little bobcats in safety on the cliff.

In the late fall or early summer of 1961, we lost Bill Brown to an undisclosed illness. With Bill's passing, Elena became even closer to us than before. Eventually their house was sold, and she moved back to Arlington. Nevertheless, we were now part of Elena's family, and our ties became even more secure through her granddaughter, Susan Dalby. We met Susan when she was about sixteen years old when she came to Tucson to visit Elena. Susan started college at George Washington University when her parents were living in the Washington area, but the lure of the desert and nature led her to Tucson and the University of Arizona where she finished her undergraduate work. We became surrogate parents for her, and she was frequently in and out of our home. We enjoyed our local natural history with her, and when she decided to pursue an MS degree in ornithology, I became an unofficial mentor. Her major professor was the ornithologist Steve Russell, a friend and colleague; and when she became interested in the migration of white-crowned sparrows, I provided her lab space and a supply of prolactin that she used in her work. After an interesting few years, Susan met Eskild Petersen, a young Danish intern at the University of Arizona College of Medicine. They were married and ultimately purchased the Bond house next door. They reared two daughters while Eskild joined the medical school faculty. Susan attended medical school, and she became a successful pediatrician. Our family friendship deepened further as neighbors. Their children have grown up and moved on, and Susan and Eskild have retired. Nevertheless, we still look toward Susan and Eskild as almost kids of our own. They are our chosen family.

Pigment cells for their own sake

Up to the early 1960s, my involvement with pigment cells and pigmentation was based upon their use as vehicles to study basic problems in developmental biology and endocrinology. In the process, I became more interested in the pigment cells themselves. I suppose that this direction was started by the discovery that the pigments of yellow or red pigment cells were pteridines. This inclination was further strengthened by my curiosity about what happened to guanophores in response to MSH. About this time, I began an investigation of the ultrastructure of pigment cells which led to the discovery that melanin-bearing cells, melanophores or melanocytes, contain a specific organelle in which the pigment was located. It became known as a melanosome. Credit for this discovery was given to Makoto Seiji, a dermatologist at Harvard Medical School, but in fact its existence was reported simultaneously by a developmental biology graduate student, Frank Moyer, who, I believe, was at John Hopkins University.

I enlisted my friend Sam Ferris to aid in studying the structure of pigment cells. We had just done a little study together at the ultrastructural level of the relationship of the optic vesicle and lens epithelium in the chick embryo. We now turned to pigment cells. One of my new graduate students, John Taylor, did much of the work. While he focused primarily on the reflecting pigment cells that we then called guanophores, we looked at xanthophores as well. John was well trained in electronmicroscopy by Sam Ferris, and he soon excelled technically. Over the next several years, John paid particular attention to the guanophores, and in fact, these cells and their unusual organelle became the subject of his PhD dissertation. He found that the reflective properties of the cells were due to the presence of purine crystals in rather flat organelles that we named reflecting platelets. These reflecting platelets were often arranged in stacks within the cell, and their precise orientation was responsible for their reflective property. In light of these observations, it was approaching time to consider a more appropriate name for guanophores, and that would soon come.

Although my attention was turning more and more toward the structure and composition of pigment cells, I had not forgotten the endocrine aspects of the previous few years. In fact, in early 1964, with the help of an undergraduate student, Sharon Procter, who was working in the lab, I did a neat little study based upon the use of MSH (or ACTH) peptides. In the late 1950s and early 1960s, several hormone chemists were working on the elucidation of the structure of ACTH. Since the first thirteen amino acids of ACTH were, in fact, identical to the MSH molecule, I was able to obtain short peptides that were identical in each. One of these was a pentapeptide, the shortest sequence capable of stimulating melanophores to disperse. I found that all of these peptides stimulated melanophore expansion and guanophore contraction in tadpoles, and that the response of the two chromatophores was always parallel. One sample of the pentapeptide contained a D-phenylalanine which was known to potentiate the chromatophore response, and indeed this was the case for both melanophores and guanophores. This study is another of my pieces of unfinished business.

Not long after my visit at Keio University, one of the young men that I met, Jiro Matsumoto, began concentrating on the bright yellow and red pigmentation of fish, notably *Xiphophorous helleri*, the swordtail fish. He later found that their bright red pteridine pigment, drosopterin, was located in particular organelles that were composed of concentric lamellae. In a classic paper published in the *Journal of Cell Biology* in 1965, Jiro described the organelle and named it a pterinosome. The swordtail chromatophore in which these pterinosomes are found are a bit more complicated, because they also contain yellow carotenoid pigments. This added to the complexity of pigment cell nomenclature. The need for clarification was in order, especially in light of the new knowledge about all the pigment cell types present in the array of vertebrate classes from fishes to man. Such was the state of affairs in the mid-1960s when I was invited, thanks to the recommendation of Howard Bern, to write an article about pigment cells for the *International Review of Cytology*. I found this to be a perfect venue for the clarification of pigment cell nomenclature.

Many aspects needed to be considered, keeping in mind historical concepts as well as the recent findings from biochemistry and electronmicroscopy. First, the two suffixes "-cyte" and "-phore" needed to be reconciled. "Cyte" (cell) was derived from the German literature of the nineteenth century that referred to -zyten. "Phore" (bearer of) had its roots from an Italian paper published in 1810 in which pigment cells were called *chromofori*. Over the years, it evolved that lower vertebrate pigment cells were called chromatophores, whereas such cells in birds and mammals were designated melanocytes. It turned out that this separation was a fortuitous happening. The pigment cells of birds and mammals are exclusively epidermal pigment cells incapable of rapid color change, whereas chromatophores are largely dermal and are the essential agents of rapid color change. One of the problems of chromatophore nomenclature was that the cells were named according to what they contained; hence guanine-containing chromatophores were called guanophores, despite the fact that other crystalline components were contained in these cells. I decided that what the chromatophores looked like was a better criterion for nomenclature. Thus, a yellow pigment cell should be called a "xanthophore" regardless of whether its yellow pigments were pteridines, carotenoids, or both. Similarly, a red-colored chromatophore should be designated as an "erythrophore" (Greek *erythros*, meaning red or reddish). The case of what used to be called a guanophore was a bit more complicated since it contained crystalline "pigments." The term "iridophore" was applied to these cells because they often appeared to be iridescent. This general scheme that I suggested has pretty well held up over the years and has even expanded. For example, in some purine-containing pigment cells, the crystalline contents are not organized in neatly arranged platelets, and thus appear white due to the scatter of all wavelengths. They are quite appropriately designated "leucophores" (Greek *leukos*, meaning white or bright). In recent years, blue pigment cells containing a true blue pigment have been described. They are known as "cyanophores" (Greek *kyanos*, meaning dark blue enamel).

During the early 1960s, I focused on these bright-colored pigment cells—iridophores, xanthophores, and erythrophores. To a certain degree this interest was due to the "Japanese influence." When I visited Tadao Hama and his colleagues in Yokohama, I issued a blanket invitation for any of them to come to my lab in Tucson. The first to visit was Hama himself, who stayed with us for a few days in 1962 *en route* to an international pigment cell meeting in New York. In fact, Professor Hama was the very first houseguest in our newly built home. When he left, he promised to send one of his protégés to my lab for a period of research.

That person was Masataka Obika who arrived in the fall of 1962 and remained for two years. I remember picking him up at the airport, and right then and there we decided that Masataka was too long, and that he needed an American diminutive name. In those days the TV serial "Gunsmoke" was in vogue, and its hero was Matt Dillon. Nothing would do other than that we call our new friend Matt. Masataka was delighted by the name and adopted the Japanese phonetic spelling Mat. Thus, for these almost fifty years he has been Mat Obika! Mat fit into the lab perfectly and became a friend for life. He was especially happy to be in Tucson, not only because of the work that we did together, but also because of his very serious hobby. Mat was a collector of long-horned beetles (cerambycids) and was famous for his vast collection of these extraordinarily beautiful beetles. He was looking forward to coming to Tucson because Madera Canyon was world renown as a collecting site for cerambycids. Because of my own interest in natural history, I encouraged him and enjoyed his hobby vicariously. We invited him to dinner frequently, and during those evenings, Mat would set up a black light and white sheet at our clothesline. For years after his return to Japan, I had a "killing bottle" and collected beetles for Mat.

During Mat's stay in my lab, I made available a large variety of amphibians, and there were many others available for collection in southern Arizona. We decided to make a survey of the varieties of pteridines present in the skin of these various amphibians at different stages, larval and adult. These data provided enough information for us to generate a rather substantial paper on the subject. Mat contributed to the lab in other ways. He was a good resource for my two PhD students, Lee Stackhouse and John Taylor, and helped augment the scientific and personal ambiance in the lab in general. Everyone interacted well with one another, including another Japanese post-doctoral student who was working in an entirely different area. He was Katsutoshi (Toshi) Imai who arrived at about the same time as Mat.

Toshi was a surprise visitor who came via Holland, rather than directly from Japan. He was a comparative endocrinologist on a two-year leave from Maebashi, Japan. He had spent a large part of his leave in Holland and was to finish his work in Howard Bern's lab in Berkeley. However, he was attracted by my recent work on the pineal and decided to spend six months with us. He got along well with the guys in the lab and was often a source of amusement. One event was rather hilarious and concerned the fact that he wanted a car while he was in Tucson. Lee was the most enterprising of the gang, and he went out with Toshi to buy a car. They ended up with an enormous old Buick, and when they returned with the car, we could hardly contain ourselves. Toshi was an extremely small person, barely five feet tall, if that. When he sat behind the wheel in his Buick he could hardly be seen! He went on to the Bern lab for his final six months before returning to Maebashi. Some time later, Howard and I were chuckling about Toshi's visit to the USA. Howard was in frequent contact with the Maebashi lab in which Toshi's wife was a prominent member. It seemed that nine months after Toshi's return to Maebashi, his wife gave birth. Howard remarked that as soon as Toshi's plane stopped on the runway, he must have jumped into bed!

7

A Sabbatical in Paris

I became eligible for a sabbatical academic year in 1962-1963. However, these were busy times. My research career was in full swing with grants and graduate students to supervise, and I did not feel that I could leave. I had been in contact with Professor Gallien in Paris, and he was willing to shepherd my candidacy for a Fulbright award to work in his lab at the University of Paris. I asked him to set the wheels in motion so that I could be in Paris for seven months beginning in January 1964. His influence was profound, and soon I was awarded a Fulbright grant. The award was generous and included cabin class accommodations from New York to Paris. In those days, there were regularly scheduled passenger ships to and from Europe. A bit of a problem was that we would need to provide passage for Lou from our own funds. Cabin class fare for her would have been very costly, but we found that I could exchange my cabin class accommodations for two places in tourist class. This was the choice we made.

We boarded the original Queen Mary in New York, found our cabin in the bowels of the ship, and away we sailed. In January, the Atlantic is not exactly a millpond, but a few days of rough seas did not put a damper on the fun we had. There was something going on constantly, and we were young and full of ourselves. Moreover, at this time of year the ship had relatively few tourists. Many of the passengers were young people, and we made many friends.

We disembarked at Cherbourg and took the boat train to Paris. Professor Gallien was at Gare St. Lazare to meet us. The picture that is embedded firmly in my mind is of Big Lou standing there wearing a dark blue overcoat and a Russian fur hat. Off we went to the Hotel California on rue des Ecoles in the Latin Quarter. We were to unpack and get settled until the late afternoon when Big Lou would return to take us to his home for dinner. This was not a surprise, for after all I knew him from Iowa City and from having seen him again at the endocrinology symposium in Japan. It was interesting, though, in view of a warning made in my Fulbright documents that I was not to expect very much personal hospitality from my host professor. Nothing could have been farther from the truth in our experience. At any rate, we were picked up and taken to the Gallien residence on rue Gazan where we met Madame Gallien (Drée), their daughter, Suzanne, who was about Lou's age, and their son, Claude, who was a university student. Big Lou and Drée spoke moderate English, but Claude spoke perfectly good American English that he learned on the streets of New York City. It seems that after Lou and I had left Iowa City, Claude was to spend a summer there with the Witschis essentially to learn English. Unfortunately, Mrs. Witschi had a recurrence of a heart problem and this precluded the possibility of hosting the young Claude. So, they sent him to Nyack, New York, to stay with their daughter Mia and her family. Claude was seventeen years old at the time, and in order to keep him entertained, Mia and her husband sent him off to the city on the train every morning. Claude loved roaming around New York City, and the experience was an ideal means to learn English. Suzanne could speak no English. In fact, she could hardly speak French. When she was a small child, she developed a neurological condition that made her profoundly deaf. Thus, she had never heard French spoken and as a result could only speak on the basis of lip reading. I'm not sure how much this was a handicap because Suzanne made a success of life. She became a dentist and practiced until her retirement about ten years ago.

The evening with the Galliens was a great success. Of course, Big Lou and Drée reminisced about their six-week stay in Iowa City and waxed enthusiastically about the kindness they were shown there. Perhaps the warm hospitality that they bestowed upon us was a reflection of their happy experiences in Iowa. During the course of the evening, a plan was set forth. We were to stay at the Hotel California until we were established in an apartment.

Big Lou had already set things in motion by assigning the job of finding housing for us to one of his *élèves* (graduate student), Marie-Thérèse Chalumeau. She had found two possibilities, one in the 11th and one in the 6th Arrondisement. We chose the more modest of the two, in the 6th, despite its rather bizarre arrangement. It was a fifth-floor apartment owned by a widow, Madame Fallex. I'm not sure that we ever knew her real first name, but she referred to herself as "Pretty." Her apartment was not large, but was rather nice. Madame Fallex's late husband was a homeopathic physician, and one room of the apartment had served as his office. When he died, Madame Fallex, his second wife, inherited the apartment, but that was about all. Because of French inheritance laws, the estate went to his son, a product of his first marriage. In order to make ends meet, Madame Fallex renovated a part of her apartment to make a small studio apartment that included a bathroom, a miniature kitchen, and one room with two windows overlooking the rue du Cherche Midi. This room was both our living room and bedroom. A larger functional kitchen with table, oven, and refrigerator was to be shared by her and us. As in most French homes and apartments in those times, the toilet was just inside the door and was again to be shared. It was all very strange, but comfortable.

Madame Fallex was very nice and could speak reasonably good English. I suspect that it was for this reason that we chose her apartment since neither one of us had ever studied French. It should be recalled that in those days the president of France was Charles de Gaulle, the heroic French General of World War II. "Le Grand Charles," as he was referred to, was extremely chauvinistic and insisted that the French language was The Language. He insisted that all contamination from other languages was to be avoided! Of course, Big Lou was a real patriot and followed de Gaulle's directive. So when we left the Galliens after our first evening with them, Big Lou announced that there would be no more English spoken! This is the way it was for the next seven months. Thus, we were essentially forced to learn to speak French. This was a cruel thing for all who had to listen to us, as we did what the French say *"casser les oreilles"* (break the ears) of the listeners. For me, it was a little easier than it was for Lou. I knew enough of southern Italian dialects to be able to understand a little, and I heard French and tried to speak it every day in the laboratory (with a slight Italian accent). On the other hand, Lou's daily exposure to French was restricted to going out to do our shopping. Of course, Madame Fallex helped her enormously. In those days, there were few, if any, supermarkets. Instead, there was a series of little stores for all of the principal staples. Like most French housewives, Lou went shopping almost every day, and thus we always had fresh food. She had her list of items written in French beforehand.

The rue du Cherche Midi is in a good neighborhood of the 6th arrondissement. This is a noted street with a long history. We were very happy with the location. There were good stores nearby, and transportation was most convenient. We were close to two Metro stations, Sevres Babylon and St. Placide. The Metro system was excellent then and is even more efficient in these more modern times. Lou sometimes used the bus, because it is much more scenic to be above ground, and bus routes are always nearby. I was pretty serious about my work, and I went to the lab every day. During the first few months, I took the Metro to the Jussieu station and I could be at work in about twenty-five minutes. Lou on the other hand had no fixed schedule and was able to take full advantage of all the cultural offerings available in Paris. She got plenty of good exercise because in Paris one walks and walks. In addition to walking, there were plenty of stairs to climb and, since our buildings did not have an elevator, Lou made the climb to our fifth-floor apartment several times a day. At the end of our seven-month stay, she had strong legs and had lost a couple of inches from her hips.

Lou's daily *commissions* (shopping) often included stops at the *épicerie* (all purpose grocer) near the door to our building. Very near it was the *boucherie* that sold excellent meat. Down a block was the *cour des halles* that offered gorgeous fruits and vegetables. The young men who ran it were nice and were most helpful. For years afterwards when we were in Paris, we made it a point to go visit them. Also extremely nice was the couple who owned the *laiterie* where we bought milk, cheese, and eggs. There were several *boulangeries* (bakeries) in our immediate neighborhood. Each one was closed a different day of the week in order that fresh bread be always available. I usually did the bread shopping and did so in the French way, which was to have a fresh *baguette*, often still warm, for the major meals. Thus, most housewives would buy freshly baked bread in the morning when they did their daily shopping. In those days, the major meal was at midday, and all the shops and stores closed from 1:00 until 4:00 or 5:00 p.m. During this interim, the *boulangers* were busy so that when they opened again at 5:00 p.m., fresh bread was available for the evening meal. When I came home from the lab, I would stop for a fresh bread to have with our evening meal. Often we did the French thing of keeping a little bread left over to make *pain grillé* (toast) for breakfast. Actually we always had fresh croissants for breakfast thanks to Sophie, Madame Fallex's *femme de ménage*. As in many established families, Madame Fallex employed a woman to do the menial household tasks. Sophie arrived at about 7:30 or 8:00 a.m. and was willing to stop at a *boulangerie en route* to buy croissants for our breakfast. These were ordinary croissants, but on Sunday mornings, when there was no Sophie, I descended to buy *croissants au beurre* (butter) for our special Sunday breakfast.

A rare photo (1964) of the centuries-old Halles aux Vins (wine market) which was demolished and replaced by the Faculty of Sciences building seen in the background.

The day after our soirée at the Galliens' was our first day in Paris. We took ourselves to the *Laboratoire d'Embryologie* where Big Lou was the *professeur*. It occupied the fifth and sixth floors of Building C of the new Faculty of Science. The building faced on the quai St. Bernard on the left bank of the Seine, just upstream from Notre Dame and Ile St. Louis. The location is the site of the old Halles aux Vins where wine was brought in from all parts of France. It was being demolished as new Faculty of Science buildings were erected, one by one. Our building, *bâtiment C*, was the first to be finished. This new "campus" extended from the left bank of the Seine to Place Jussieu and the Metro stop that I used coming and going to work every day. It was special to traverse the old Halles aux Vins twice a day as I thought of all the wine that was brought here over the centuries from all parts of France.

That first day of initiation into living in Paris, we went to lunch with Big Lou at a bar next to the entry to the Jussieu Metro stop. Bars in France are not at all like bars in the USA. First of all, it is likely that more coffee and soft drinks are served than alcoholic beverages, although plenty of beer and wine are consumed. Most bars serve foods other than just sandwiches. Many serve hot meals and especially a *plat du jour*, the special of the day. Big Lou wished to acquaint us with the latter, and he pointed out that such meals are typical and quite reasonably priced. We never developed the French habit of having a major meal at noon, so we did not often take advantage of the *plat du jour*. However, a few days after our experience with Big Lou, we found ourselves near that same bar at mid-day. Lou decided that she would have their *plat du jour*, which happened to be *tripes á la mode de Caen* (tripe in the manner of Caen, a small city in Normandy). She had never eaten tripe, but I had been forced to eat it as a boy, and for sure I would never eat it voluntarily. That was Lou's one and only experience with tripe as she pointed out most emphatically.

Our daily life in Paris was like that of most French families. I went to work, or rather to the lab, almost every weekday, and Lou fended for herself. There was much to keep her busy. I took the Metro about 9:00 a.m. and returned about 5:30 p.m. It was an historic line that took me under the Latin Quarter where every Metro stop had its history. Before we left Tucson, we had ordered a new VW beetle for pickup at the factory in Wolfsburg. No specific date for the pickup had been established. After our arrival in Paris when I saw the traffic and the scarcity of parking, I was in no hurry to go to Wolfsburg. Moreover, public transportation in Paris was superb and for the first month or so we had no desire to leave on weekend trips. We did as most Parisian families did by taking advantage of the parks and monuments. It was really a joy on Sundays to see families out strolling with their little kids in tow. Then, the weekend was over and everyone resumed their schedules. I actually enjoyed going back to work because the lab was a pleasant place, and the people there, service people and graduate students alike, were excellent and most kind.

This was a prosperous time in France as recovery from WWII was still in full swing. Professor Gallien's laboratory was well equipped with first-rate instruments. Most of the offices and individual laboratories were on the fifth floor and my particular "*pièce*" (room) overlooked the Seine. I could not have been in a better spot for I could look down the river to see the barges going by, and by looking off in other directions, the major monuments came into view. The very end of the corridor terminated in a large, fully glassed room, essentially an aquarium that housed breeding colonies of a variety of amphibians. The view that it provided was spectacular—there right ahead was the apse of Notre Dame and to one side on the top floor of a nearby building was the famous Tour d'Argent restaurant. Unfortunately, about twenty years ago the view was blocked by the erection of a new building dedicated to Moslem culture. The magnificent view is gone, but it is etched indelibly in our memories.

That Professor Gallien could have such a fully equipped facility at his disposal, and all the funds needed to operate it, was a function of the times. He occupied the Chair of Professor of Embryology at the University of Paris. There were no other professors, and as well as I can recall, there was not even a *Maître de Conférence* (es-

A young Claude Gallien on the roof of Batiment C in which the Laboratoire d'Embrologie was located. The apse of Notre Dame can be seen in the background.

sentially Associate Professor). There were also about ten or twelve *élèves* (advanced graduate students) to help in the teaching laboratories. Similar chairs in other disciplines such as physiology, anatomy, and so on, were also occupied by other individual professors. This was a system that all came tumbling down a few years later when the "events" of May 1968 brought a huge reform. The many sometimes violent demonstrations based upon the lack of adequate university positions for young people led to the establishment of new universities and the creation of new professorships. We happened to be in Paris again for two months during the summer of 1968 when I was finishing some work in Big Lou's lab. It was a time when the administrative changes were just underway and severe security measures were in place. All of these changes of course did not make the Gallien family very happy. I decided to not get myself involved in discussions about them. Instead, I devoted my energies to the experimental work at hand. In any case, there were no members of the Gallien family around since they were all in St. Briac in Brittany where they had a summer home.

In 1964, we did not see much evidence of possible unrest although one of the *élèves* was said to have become a bit rebellious, and Big Lou asked him to leave. The remaining *élèves* seemed to be pretty calm, although they all had their own strong characters and opinions. I got to know them all quite well, and Lou and I were invited to their respective homes during the course of our stay. We went together to the "canteen" of the faculty for our midday meal. It was cafeteria style, but was typically French. The offerings included some sort of a salad, such as beets, finely grated carrots or celery root, or Bibb lettuce, together with a meat course. Dessert often consisted of a wedge of Camembert or a pot of yoghurt. Afterwards, we would leave the building to walk to a nearby corner at the junction of boulevard St. Germain and quai St. Bernard where there was a fine bar. I'll never forget its name, "*J'ai du bon tabac*" (I have good tobacco.). We each had our cup of French espresso and engaged in lively conversation before our return to the lab for our afternoon's work. It was such social interactions that helped me to learn to speak French.

Getting myself set up to do my experiments was made much easier thanks to our friend Marie-Thérèse Chalumeau. It was she who found the many small instruments that I needed for embryonic microsurgery. I still have the small metal recycled cookie box that has her name on it. She also explained how the lab was run and how to obtain embryos for my work. She had been a long-time member of the Gallien team and had been part of the move that brought the lab from their old building on rue Cuvier to the new *bâtiment C*. When she learned that I had been trained by Hans Holtfreter at Rochester, she regaled me with tales of his exploits in Paris.

Some years earlier, probably during my early years at Arizona, Big Lou had invited Holtfreter to spend time in their lab on rue Cuvier. There, he very much impressed the Gallien laboratory. Of course they were all in awe to have this famous scientist among them, but to have him do his microsurgery using sterile techniques on one hand while he had a lighted cigarette dangling from his mouth was amazing. A souvenir of the Holtfreter visit hung in the library of the laboratory. It was a print of a black and white, pen and ink drawing that Holtfreter had drafted some years earlier. It was kind of an allegorical view of evolution. I am happy to own a copy of that print that Hans gave me a few years before his death.

During my tenure in the Gallien lab, I worked on a variety of projects. I was free to do whatever I wished. One project that turned out to be significant was based upon the unique presence in the lab of a breeding colony of Algerian newts, *Pleurodeles waltlii*. This species is native to Algeria and Tunisia, and when brought to France, it was discovered that they were easy to breed. Moreover, they produced beautiful embryos that were particularly amenable to microsurgery. Most of the Gallien team used them for their research. I had never seen the species and in fact had little experience with newts. Newts belong to the Salamandrid family, and there are many species in Europe. When Mat and I did our survey of integumental pteridine distribution among amphibians, we had included the Japanese newt, but we did not have any European species to test. There were a few of these available in Big Lou's lab, and others could be caught in ponds near Paris.

The first newt species I tested was *Pleurodeles*, and to my surprise the fluorescence pattern of its pteridines differed from that of other salamanders. Moreover, with paper chromatography, there was one pteridine that migrated faster than the others, and it emitted a characteristic blue fluorescence. For the sake of convenience, I began referring to this substance as "pleurodeles blue." One of the questions that came to mind about pleurodeles blue was how did it relate to developmental stage. This was quite an appropriate question because Big Lou and one of his associates had some years earlier devised a staging series for *Pleurodeles*. These embryos were so commonly used among the Gallien group that it was important to assign numbers to specific stages of development. I was able to ascertain when pleurodeles blue first appeared, but more important, I learned that it disappeared abruptly at a specific point during metamorphosis. This recalled what Mat and I had learned about differences between adults and larvae of other species. Since *Pleurodeles* embryos and larvae of all stages were now available in number, I decided to investigate the cause. It was logical to implicate thyroxine as a cause for the disappearance of pleurodeles blue since not only was it known as the hormone that brings about amphibian metamorphosis, but also each specific event in metamorphosis possesses a specific level of sensitivity to the hormone. Jane Kaltenbach of Mt. Holyoke College and a friend of ours had demonstrated this fact by implanting small pellets containing a given percentage of thyroxine near the site of change. I wrote to Jane and asked for some of these thyroxine/cholesterol pellets, and she kindly obliged.

My plan was to implant a pellet in one of five sites at the base of the dorsal fin of a *Pleurodeles* larva that was sufficiently large, but at a developmental stage sufficiently removed from its own metamorphosis. The plan was to observe the experimental larvae daily to check for signs of metamorphosis, viz., local resorption of the tail fin or localized molting. When such signs were evident, a small piece of skin was removed from each of the five sites and was squashed on a piece of chromatography paper. The edges of the sheet were clipped together to form a roll, and that was stood for five hours or so in a petri dish containing a solvent. As a front of solvent ascended by capillary action, it passed through the squashed skin samples carrying with it the pteridines that it had extracted. When the chromatogram was dry, it was viewed under ultraviolet illumination, and the respective pteridines revealed their presence by presenting a spot of fluorescence. The experiment was a remarkable success and revealed that pleurodeles blue was found only in the squashes from non-thyroxine exposed skin. Without fail, there was no pleurodeles blue found in the samples from the thyroxine pellet areas. These data provided the basis for a small manuscript that I wrote about the role of thyroxine in bringing about the loss of pleurodeles blue. It was written in French with considerable help from one of the élèves, René Ozon. Perhaps more properly I should say that René translated my

English version into French. Big Lou apparently was pleased with the work and thus introduced it for publication in the proceedings of the French Academy of Sciences (CR. Acad. Sc. Paris, 15 June 1964). It was presented by M. Robert Courrier, a member of the French Academy. Of course, I was pleased and honored. Some years later, I was even more pleased when Claude Gallien used the pleurodeles blue story as a basis for his doctoral thesis.

In the fall when we returned to Tucson, I used my discovery of pleurodeles blue to answer some embryological questions related to the autonomy of neural crest expression. Of course, it had long been known that pigment cells take their origin from the neural crest from whence they migrate great distances to pigment the entire animal. An unanswered question was: does the neural crest cell have an intrinsic capacity to express its characteristics without other contributions from the host? For example, is the specific expression of pteridine pigments intrinsic to the neural crest cell? I found that this is the case. In experiments in which pieces of neural crest were exchanged between *Pleurodeles* embryos and those of an unrelated salamander species, the Mexican axolotl (*Ambystoma mexicanum*), pleurodeles blue was found on axolotl hosts, and the axolotl pteridine pattern was expressed on *Pleurodeles* hosts. These experiments were done in collaboration with the late Christina Richards who worked with me while in Tucson for the academic year 1964-1965 during her husband's sabbatical. Before going further, I point out that the pleurodeles blue story is another example of unfinished business, since to this day it is not known what the role of thyroxine is in affecting integumental pteridine expression. It would have been perhaps an even greater sin if I had not again gotten involved some twenty-five years later to show, with the help of Japanese colleagues, that pleurodeles blue is not really a pteridine. Rather it is a novel pyridine derivative. In a sense, this just begs the question, for we still do not know how its metabolism is affected by thyroxine and how this relates to pteridine metabolism in general.

I return to our 1964 sojourn and life in Paris. After living there for many weeks without our own means of transportation, the idea of going to pick up our Volkswagen became more attractive. Accordingly, we made a date to be at the factory to claim our purchase and to have a tour of the assembly line. We went to Wolfsburg by train, and at the factory found it a fascinating experience to watch the basic chassis of the car move forward on the assembly line while the successive parts to be added moved above in orbits of their own. At the proper time, a part would descend to the line to be put in place, and so it was with part after part until a complete Volkswagen Beetle emerged at the end of the line. After all that entertainment and $1200.00, we had a new Beetle to drive away. We drove south and up the Rhine toward Alsace and our first required service. We crossed over to Bouxwiller where we visited Odette and her mother. The countryside was still covered with snow. I remember clearly that, as we crossed little streams, spring was coming, for the pussy willows were out as a first harbinger.

While it was not essential for us to have our own car in Paris, it was wonderful for little trips away from town. Parking on the street in our neighborhood was no real problem as long as we remembered to park our car on the correct side of the street. I believe that in those days one was to park on one side of the street for the first fifteen days and on the other side for the rest of the month. As we reminisce about the past, we laugh as Lou points out that those days have been long since gone. Now, one parks on both sides of the street wherever there is a place or even part of a place. I did not often drive to the lab, although parking was always available there. When I did take the car to work, it was usually because I had plans to play handball.

Handball was a passion of mine from the spring of 1957 until 2009 when I finally retired from the sport. When we went to Paris, I felt certain that, while the French do not play handball, I would be able to find courts and players with the American military at NATO. In fact, in those days, SHAPE (Supreme Headquarters Allied Powers in Europe) was located near Versailles. I called the athletic director at the American base. He kindly provided directions for me to get there, and we made a rendezvous at NATO headquarters. He provided me with a pass to a nearby camp where the courts were located. I made the three or four hour round trip by Metro and bus about twice a week for a bit, but once we had our Beetle, I could drive to the facility in about twenty-five minutes. I had learned about a back way to Versailles, the route des Gardes, originally the private road taken by Marie Antoinette's personal troop to travel between Paris and the Versailles palace.

Obviously, we adored our life in Paris. I did spend long hours in the lab, during which time Lou enjoyed a total immersion in the culture, but we also had a unique social life. We did many things with the young people in the lab and their families, but most important was the interaction with the Gallien family. We had dinner at their home almost every Thursday evening. Their home, on rue Gazan, overlooked Parc de Montsouris which also bordered on the boulevard Jourdan. On the opposite side of the boulevard is Cité Universitaire, a long string of university buildings named after countries, viz., Maison d'Angleterre, Maison de Danemark, and so on that extends from Porte de Gentilly almost to Porte d'Orleans. On the corner, diagonally across from Parc de Montsouris and bordered by Avenue de la Porte de Gentilly, is PUC (Paris University Club) and Charléty Stadium. Lou would arrive by Metro or bus, and I would walk over from PUC with Claude Gallien. On Thursday afternoons, Claude, who was still a student, and I would take the Metro to Cité Universitaire, and walk over to PUC, where I would play basketball, and Claude would do a weight workout.

It was perfect to end the day with dinner at the Galliens' who treated us like family. They seemed to be pleased when Lou invited them to our miniscule apartment for dinner. I remember well that she prepared an excellent pork roast, and that dinner was a grand success; however, I just don't remember where we found places for everyone to sit! As further testimony to our friendship, the Galliens invited us to spend a weekend with them at their vacation home in St. Briac on the coast of Brittany. Some years earlier, they had purchased and restored the home of a retired sea captain. It is a beautiful place and has a lovely garden. We had great fun there with all the family, including their dog PUC, a smallish kind of French poodle that is called a *caniche* in France. She knew us well from Thursday evenings in Paris.

And so the seven months in France passed by altogether too fast. We were comfortable and at home, and in fact, we could have easily stayed to make our life there as expatriots. Perhaps our experience was unique, but on the other hand, we may have had a different point of view. For my part, I never sensed in the French what many American visitors took to be arrogance and impatience. We felt that Parisiens were no different than big city dwellers elsewhere in the world. In the country, the folks we met were friendly and calm as they put up with our attempts to speak their language. As time went on, we both made progress in this regard. I was especially proud of Lou's success considering that she did not come from a romance language background. My use of the language was adequate enough for Big Lou to ask me to give a seminar on my work. Moreover, he must have been sufficiently pleased to offer me the position of *maître de conférence*, essentially equivalent to our rank of associate professor. Of course I felt honored and proud at the offer. I was able to decline gracefully since I had just learned that I was to be promoted to full professor at the University of Arizona. I was proud of this promotion, since I was still in my mid-thirties. The day too soon arrived when we bade farewell to France and boarded the Holland American *Rotterdam* and sailed back home.

It was not long before it became possible to repay some of the hospitality that we were shown by the Galliens when, about a year later, Big Lou came to Baltimore for an international conference. We were pleased when he accepted our invitation to visit us in Tucson as part of his trip. His stay with us has provided many memories and quotations. It was in early September and still the monsoon season. Of course it rained, which prompted Big Lou to proclaim, "*C'est un desert paradoxique—il pleut!*" (It is a paradoxical desert—it rains!) We had a wonderful time showing him around, but because it was the start of the school year, it was hard for me to get away. Since I did not have time to take him to the Grand Canyon, we booked him on a bus tour, from Tucson to the Grand Canyon and return. It worked out beautifully, and he was pleased. However, when he got back, he informed us that mules in Arizona are very weak! He had hoped to take a mule ride to the bottom of the canyon, but he was denied this opportunity because there was a weight limitation imposed upon riders. Big Lou was a big man—not fat, just big—and his weight of a bit over 200 pounds exceeded the limit! We were sorry, but at least it gave him one more story to tell when he returned home.

Our friendship with the Gallien family has deepened over these almost fifty years since my 1964 sabbatical. We have visited them in Paris many times, and we have had the pleasure of a visit to us in Tucson from Claude and his wife, Michèle. Unfortunately, Big Lou passed away earlier than he should have from an unexpected heart attack, but Drée remained at the rue Gazan apartment. When we visited, Suzanne, Claude and Michèle, and their family would gather there for an afternoon or evening reunion and a champagne toast. After Drée passed on, the reunions were at Claude and Michèle's apartment just outside of Paris. These Gallien gatherings were

Claude Gallien's cartoon.

larger than before since the Gallien granddaughters, Caroline and Barbara, and their families were also there to greet us. There was much to talk about at these gatherings, as we needed to be brought up to date about what had transpired since Claude's annual year-end holiday message.

There was always much going on in Claude's life for, aside from his teaching as a professor of embryology, he was an active author, having written several textbooks. I believe that he has given me inscribed copies of all of them. His life in international sports is even more impressive now that he has retired from university life. He was recently installed as president of the International University Sports Federation (FISU) after an election in which he defeated the often-controversial John Killian, an ancient American who seemed to have been president forever! Claude's ascension to this position seems an almost natural progression from his life-long involvement with PUC and its stadium, Charlety. In fact, he was President of PUC for about fifteen years.

Among other books that Claude wrote were histories of Charlety and of PUC. At one of our gatherings of the Gallien clan, Claude presented me a copy of the Charlety book. I noticed that it had not been inscribed, so he retired to a corner with book and pen in hand. After several minutes, he returned with the book, and I saw his inscription. It was, in fact, a cartoon (Claude is a talented cartoonist) based upon an event that occurred at PUC back in 1964 on one of those Thursday evenings that preceded dinner at the Galliens'. When we played basketball back then, the group of guys was pretty much the same, and we all knew one another. On this particular evening, there were two young women, identical twins, who were quite good players. They were members of the French women's national team. They were tough looking babes, and I said to myself, "I bet their father was a prison guard at Auschwitz!" I was probably one of the best of the men playing, and perhaps that was why one of the twins guarded me. Right off, she started pushing, grabbing, and hacking away at me. I put up with it as long as I could, until finally, I pinched her on the butt. She shrieked, jumped, and continued playing, but all the physical play ceased. Claude's cartoon recorded it perfectly, even to the authentic caricature of yours truly.

8
Some Productive Years

There was much catching up to do upon my return. The teaching load was heavy, especially in comparison with what is expected from faculty nowadays. Even then, promotion to full professor was greeted by some as an invitation to slow down and be comfortable. Having achieved tenure was their goal, and with the security it brought, life was good. Frankly, I never thought about it, because the mindset that I was blessed with was like that of my friends at Berkeley. Howard Bern always told me that his colleagues at the University of California just worked hard at both teaching and research because they wanted to. Those who did not toe this line could sense the pride held by their fellow faculty members and just left without pressure from the administration. My publication record began to mount, and with it was recognition at the national and international levels.

A Bulgarian experience

Not long after our return from France, I was invited to speak at what was essentially a melanoma conference to be held in Sofia, Bulgaria, in 1965. It was the VIth International Pigment Cell Conference and was sponsored by the International Union Against Cancer. To host such a prestigious international conference was quite an accomplishment for the Bulgarians. This conference was one of the more memorable scientific meetings of my career. The travel award to pay for my participation was not great, so I took advantage of a considerably reduced airfare if I left a day or two earlier in May than necessary. Thus, I arrived a bit early. It was a bit complicated to fly to Sofia. I needed to go by way of Vienna, from which there was only one flight per week. This was the height of the "Cold War," and thus there was little travel to and from the west.

With my early arrival, I planned to do whatever sightseeing I could before the conference. I stayed in the only hotel available to non-communist guests. It was comfortable enough and the dining room was interesting, but not elegant. The first evening I encountered a fellow American who was an official with the US Embassy in Bonn, Germany. He told me that he and I were the only Americans in Sofia at that moment. His was to be only a brief stay since he was there only to exchange maps with the Bulgarian authorities. It seems that he made frequent trips behind the Iron Curtain and was aware that Americans were under close surveillance. He warned me to take great care and not to put myself in a compromising position that could by exploited by the communists. Having had the fear of God put into me, I took great care to follow his advice, except for a social interaction that arose with some Bulgarian college students.

There were a few English-speaking students who seemed to be around the hotel entrance looking for the opportunity to speak with Americans. They had a surprising knowledge about the USA. I had a pleasant conversation with one young man, a medical student whose physique told me that he was also a weight lifter. He asked me all about American weight lifters, a sport about which I knew little. He took out a well-worn clipping from his wallet to show me photos of lifters he knew about. These students seemed to have some access to American magazines. This source, together with listening to Radio Free Europe, probably accounted for their knowledge of English. A group of these young people gathered around me and asked if I wanted to go with them to the state-operated nightclub. So off we went by public streetcar. More than likely, they asked me to be with them because the club sold some merchandise from the west that could only be purchased with dollars. We had a great time, although I'm sure that the official from Bonn would have been aghast if he knew what I had done.

Language was a problem for most of the attendees of the conference, because the number of Bulgarian hosts who could speak English was few indeed. Aside from Bulgarian and Russian, most of the educated Bulgarians spoke at least one other foreign language dependent upon their age group. Older people spoke German, while a younger age group spoke French.

By the last day of May, a few more conference attendees began to arrive. The first of the Americans was Walt Quevedo from Brown University. He and I were invited to a parade and to view it from a special reviewing stand along the route. This year's parade was especially important since it honored the 1000th anniversary of the birth of the Cyrillic alphabet. Aside from marching groups, many of them students, there was a military element represented, mostly by heavy weapons. Most impressive of all were the many floats, many of them anti-American, designed to protest the American involvement in Viet Nam. I don't remember how Walt felt about these, but I was amused and interested.

In our viewing group were several Soviet participants at our conference. We were engaged in conversation by a large Russian woman who was a surgeon in Moscow. With her was a smallish man who bore Asian facial features; he was a physician and melanoma researcher from Alma Ata, now Almaty, the capital of Kazakhstan. They both spoke English well and were very pleasant. We walked toward our hotel together. *En route*, we learned that their rooms were in a better hotel than ours. In fact, since they were Communist Party members, they were accommodated in a quite nice hotel. They wished to extend the conversation and invited us to see their hotel. Our friend from Alma Ata asked if we would all come to his room to sample the caviar that he had brought with him. His uncle was from the Caspian Sea area and fished for sturgeon. When he stopped to visit *en route* to Sofia, he was presented a large plastic bag filled with caviar. Of course, Walt and I enjoyed the caviar and crackers that we washed down with *slivova (slivovitsa)*, a special plum brandy whose interesting history dates back to the fourteenth century. It was brought by the Russian woman. This pleasant social interaction was a fine way to start the conference.

I was among the first speakers the day the conference began. The morning was devoted to the basics of pigmentation. We adjourned to our hotel for lunch, and as I entered the conference hall upon our return, I was stopped by one of the Bulgarian staff members who spoke little or no English. To my surprise, he said clearly, "Professor Bagnara, *shönes mädchen will mit sie sprechen.*" I certainly understood that a beautiful girl wanted to speak with me, so I followed him to where, indeed, a beautiful young woman was standing. She was introduced as Kana Trendofilova, a laboratory technician at the Cancer Institute and a conference attendee.

Her alternative language was French, so we were able to talk with one another. She wished to participate in the social events associated with the conference, but needed an escort. A thousand alarms went off in my mind as I thought of the words of admonition offered earlier by the man from Bonn. I was in a bit of a bad spot, but I consented to do so with the caveat that one of my American friends would often accompany us. I was referring to Irv Geschwind, a friend from the University of California Davis. Thus, I was Kana's escort, but I called on Irv to be with us when I felt the need. She was indeed a lovely person with whom I spent enough time to understand the desire that she and some of her friends had wanted to escape Bulgaria and the communist regime. Before I left, she presented me with several gifts, and much to my surprise, she appeared at the airport to say goodbye. After my return to Tucson, I received letters from her from time to time. I never learned if she was able to get out of Bulgaria, as did some of her friends.

A part of the conference was devoted to clinical aspects of pigmentation, so some other of my American friends and I decided to play hooky one afternoon. Earlier, I had found that there was a travel agency in the hotel and that we had the common language of French. Thus, I organized a van trip to Rila Monastery located high in the mountains not too far from Sofia. With me were Ron and Barbara Novales and Jean and Jim Burnett. Rila Monastery was a relic of the Ottoman occupation of Bulgaria. The Eastern Orthodox Church was heavily persecuted by the Turks who did not allow conventional churches in Sofia. A group of priests fled to the mountains and erected this beautiful church and monastery. The arrival of our van was a great curiosity, and we were immediately surrounded by young people who were happy to have their pictures taken. They were used to conventional cameras, but were completely bewildered by the Burnetts who had a Polaroid camera. They were amazed to have their pictures taken and then to see themselves immediately when the photos were pulled out of the camera.

One afternoon, Irv Geschwind asked me to go with him to see the synagogue of Sofia, a short walk from our hotel. We were greeted by the rabbi with whom we had difficulty communicating. What worked was that the rabbi spoke Yiddish, and both Irv and I had some facility with German. I believe also that Irv had heard Yiddish spoken when he was a boy. Anyway, we learned that during the Holocaust most of the Jews of Sofia were lost and that his congregation was now only about fifty. It was a sad, but interesting, experience.

When the Sofia conference ended, I needed to hurry home to fulfill another obligation. As I mentioned earlier, the 1960s were golden years with respect to science funding. One of my scientific acquaintances was Professor Donald Farner who was heavily engaged in avian physiology. He had a strong program in the Zoology Department of Washington State University in Pullman where he was chairman. He and his colleagues were investigating the physiological basis for bird migration, and their research organism was the white-crowned sparrow that summered in the nearby canyons of the Snake River. A few years earlier, Don was successful in obtaining a training grant in comparative physiology that funded a graduate course each summer. The summer of 1965, the course was to be in comparative endocrinology, and some months earlier Don asked me to come to Pullman to teach it. I was delighted at the prospect and accepted. I was allowed to set my own schedule provided that I gave three lectures per week. There was no problem with my selecting to teach on Tuesdays, Wednesdays, and Thursdays. Lou and I were then free to enjoy long weekends. There was much to see in eastern Washington and in Idaho.

We had the good fortune of being able to rent the small apartment of an English professor who was away for the summer. It was a great place to live and a perfect time of year. Outside the back door were trees in which there were two nests. There was a yellow warbler in one and a cedar waxwing in the other. Over the early years of our marriage, Lou tolerated my interest in birds, but our stay in Pullman whetted her own interest in this regard. First, we had these two beautiful species nesting under our noses, and second, we had such good bird habitat nearby. Pullman is situated in the Palouse country of eastern Washington. Palouse means lawn in French, and indeed these broad expanses of wheat fields appear as never-ending lawns. They are broken every now and then by discrete buttes erupting from the plain. They were covered with pine trees. Kamiak Butte, not far from Pullman, was especially pretty and had a lovely picnic area. Our favorite table was near a water spigot that dripped into a puddle that attracted many birds. Sitting there with our picnic gave us the opportunity to see several beautiful species clearly. I believe it was a gorgeous male western tanager that finally seduced Lou into becoming a novice birder.

Our return to Tucson for the fall semester meant a return to teaching. The lab was a little quieter since Lee Stackhouse had finished his PhD and was then a postdoc in Reed Flickenger's lab in Buffalo. Most of my research activity was directed toward the ultrastructure of chromatophores, and that involved the collaboration of Sam Ferris and my graduate student John Taylor. Sam had his own teaching to contend with, so John was the principal worker. That fall, I was introduced to the Mexican leaf frog, not the frog itself, but its eggs. One of the departmental graduate students who was working in ecology was assigned to assist in teaching a laboratory section of my embryology course. Early on he returned from a trip to Alamos, Sonora, Mexico, and brought with him a recently laid clutch of leaf frog embryos. They were amazing, somewhat larger than leopard frog eggs, and bright green. Little did I know that in a few years, a large part of my work would involve this species.

The dermal chromatophore unit

August of 1966 brought a new member to the lab. He was Mac Hadley, a new postdoctoral student who had just completed his PhD at Brown University under the tutelage of Walt Quevedo. I don't recall Walt having mentioned him when we were in Bulgaria together, but I knew of Mac from some correspondence I had with him in 1963 when he was just starting his PhD work. I didn't hear from him again until he applied for a USPHS postdoctoral fellowship to work with me. He started in 1966 and later obtained a renewal to continue in the lab for a second year. Mac's PhD dissertation was a large piece of work, and while it revealed no new major discoveries, it was a thorough description of the leopard frog integument with major emphasis on pigment cells. Perhaps its most important contribution was that the frog epidermis contains epidermal melanophores identical in form, function, and hormonal sensitivity to the epidermal melanocytes of birds and mammals, including humans. He also confirmed in adult frogs my earlier discoveries on tadpoles that iridophores (guanophores) are controlled by MSH. I refer to them as iridophores at this point because 1966 was when I published my review article on chromatophore nomenclature. Mac was particularly interested in endocrinology, and that was another reason that he chose to come to work with me.

When Mac and his wife Trudy arrived, we were doubly surprised. First, Mac was a bit older than the usual new PhD—in fact he was only a year or so younger than I. In addition, Trudy was well along in a pregnancy. Ultimately, Trudy gave birth to Martha, their only child. Lou and I very much liked Trudy, and over the years we have cherished our friendship. I respected Mac's work ethic and tried to help as best I could. He wished to use the frog skin bioassay method that was developed in Tom Fitzpatrick's lab at Harvard for some of his work, so we bought the basic tools, and he started work. It was not the sort of work that I wanted to do, because I hate to kill frogs. The frog skin bioassay required that the test animals be beheaded in a guillotine-like apparatus and pieces of skin removed to be treated in various hormone solutions. Responses to the test solutions were measured photometrically. I pointed out that the responses he measured were the sum of the effects of the three basic chromatophores present in that skin, and the effect on at least melanophores and iridophores needed to be elucidated. This resulted in a study that we published in *Endocrinology* (Hadley and Bagnara, 1969).

In reading Mac's dissertation, I observed that he had cited work done by German workers years ago. They had found that dendritic processes from adult frog melanophores could be seen in the dermis making contact with the surface of nearby iridophores. I suggested that these contacts between chromatophores might be synapse-like and perhaps might be a means of communication. Accordingly, I put John Taylor to work to examine this possibility at the electron microscope (EM) level. He was already looking at iridophores of the canyon tree frog (*H. arenicolor*) for his dissertation research, so I provided him with other frogs, notably the green tree frog (*H. cinerea*), to extend his study. In all these species there was indeed a contact between iridophore surfaces and melanophore processes, but the contact was passive and in no way synaptic. By looking at thick sections at the light microscope level, thin sections at the EM level, and whole skin mounts, we realized that we were observing a highly organized distribution of dermal chromatophores that functioned as a unit. We called this the "Dermal Chromatophore Unit" and published its description (Bagnara, Taylor, and Hadley, 1968) in the *Journal of Cell Biology*. It is a classic paper, one of the high points of my career.

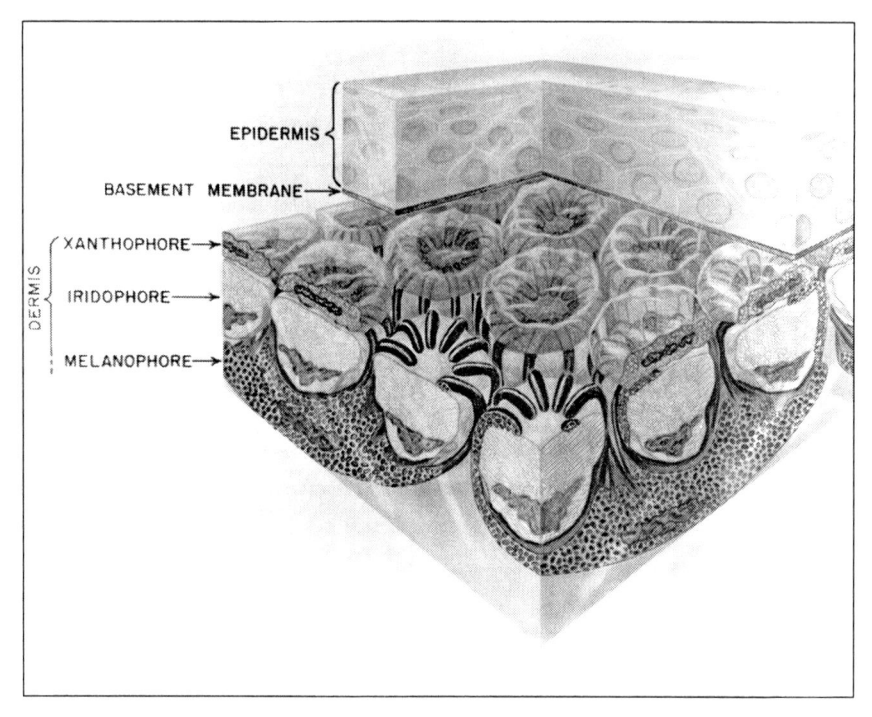

The "Dermal Chromatophore Unit" that we described in 1968.

Although I had a high regard for Mac as a scientist, I found him rather bombastic and impatient. Fortunately he had enough redeeming qualities to be forgiven by many he offended. Mac was ready for a faculty position after two years of productivity as a postdoc, and since he wished to teach endocrinology, I tried to help him find a position. There was to be a regional meeting in comparative endocrinology to be held at Friday Harbor toward the end of his second year. I recommended that he be invited. I needed to be at home and did not attend. At least two of my friends who would be in attendance were at institutions that had openings for a comparative endocrinologist, so I hoped that Mac might be recruited. When he was not, I was not surprised and thought that their loss might be our gain. I was frankly tired of the burden of representing both developmental biology and endocrinology, so I strongly urged that Mac be hired for one of our two open positions. My dear friend Newell Younggren, our department head, had already been impressed with Mac's teaching qualifications, so Mac was hired as assistant professor.

I was pleased to now to have a colleague who was also a pigment cell person, and I tried to help Mac as best I could. All the instruments that we purchased for Mac's postdoctoral work were turned over to him as he got his own lab setup. I also shared the wealth with him with respect to some writing projects that I was asked to carry out. The first was a review of the pineal for the *American Zoologist*. About the same time, Howard Bern asked me to write a small book on pigmentation as part of a series he was organizing and editing for Prentice Hall. I asked Mac if he would like to co-author it with me. Thus, in 1973, *Chromatophores and Color Change* was published and enjoyed good success. Aside from these projects, we pretty well went our separate ways. For my part, I realized that Mac was searching for his own identity and that he needed to be disconnected from me, so minimal interaction between the two of us was probably for the better in view of our respective personalities.

Over the years, as Mac's career blossomed and his sense of self-confidence rose, we interacted a little more often. In large measure, that had to do with my great respect and affection for Trudy. Mac and I never again collaborated, but we did work together. The most notable example was in 1986 when I organized and was chairman of the XIIIth International Pigment Cell Conference held in Tucson. He did yeoman work as a member of the local committee. For many years, Mac collaborated with Victor Hruby, an organic chemist, whose lab synthesized various MSH peptides. Mac's role was to test the compounds using the frog skin bioassay while other collaborators in the medical school tested other effects on humans. The goal was to obtain human applications such as sun tanning. There was much publicity associated with these projects, and Mac very enjoyed the notoriety and financial security the work provided. He wrote a successful endocrinology text that also augmented his income and reputation and facilitated his role as patriarch of the Hadley clan.

This was the state of affairs, even after Mac retired. It all ended catastrophically one afternoon in November 2006 when he returned home from a meeting and surprised an intruder in his home. The confrontation resulted in his being shot fatally. Thus, in a quick moment, this successful scientist and father was murdered. We all mourned his passing and grieved for Trudy and Martha and her children. We notified Jiro and Akiko Matsumoto who were good friends of Mac and Trudy, and they were of course shocked and saddened. When Jiro came to my lab in 1968, he and his family arrived while Lou and I were away. Mac and Trudy met them and their two young children at the airport when they arrived and the two families quickly developed a friendship. Martha Hadley was only a year old at the time. Prosecution of Mac's murderer dragged on interminably, but finally, in April 2010, he was found guilty. Jiro and Akiko made a quick three-day trip at that time. They stayed with us, and we took them to see Trudy twice. Trudy had not been well since Mac's death, and was still too weak to support long visits, but she was especially moved by Jiro and Akiko's making the special trip from Japan. Lou and I visited Trudy not so long ago and were so happy to see her well, happy, and surrounded by her loving family.

The eventful year, 1967

It is amazing how much took place in our lives during 1967. My research and teaching went well during that year, but we also had our first major health problem. The start of the year was smooth enough, especially with John Taylor finishing up all his PhD work and his having received a fellowship to finance a stay at Keio University with our friends Mat Obika and Jiro Matsumoto. I was busy writing some papers about work that I had done with Mat when he was here. The paper that Chris Richards and I wrote about pleurodeles blue was also finished. I thought that the summer of 1967 would provide a good opportunity to follow up on this work while we spent six weeks in the cool Grand Teton Mountains. To this end, I applied for and was given space at the Jackson Hole Research Station. This small facility was funded by the Rockefeller family but was operated by the University of Wyoming, Department of Zoology. It provided adequate research space and the living accommodations looked interesting.

In the spring before we were to leave, Lou had her annual checkup with our family physician. He was a bit concerned about a nodule that he felt on Lou's thyroid gland, and a follow up biopsy revealed a carcinoma of the thyroid. We were stunned, but were encouraged after the surgery by a good prognosis. The trip to Jackson Hole was delayed, so we were able to spend only three weeks of the planned six at the station. Lou recovered quickly, and only three weeks after her surgery we went on a five-mile hike. The short stay was a success in two ways. First, I was able to show that pleurodeles blue would endure in pleurodeles larval skin transplanted onto a tiger salamander host. Secondly, we enjoyed beautiful hikes and scenery in the

Jackson Hole area. As a fringe benefit, I gave up pipe smoking during the stay and have never again smoked.

After our return, it was time to make plans to participate in the Vth International Symposium on Comparative Endocrinology to be held in November in New Delhi, India, where I was to speak on the regulation of bright-colored pigment cells. Given the scare we just had over Lou's thyroid cancer, there was no question other than that she make the trip with me. Because India is on the other side of the world, we found that we did not need to go and return by the same route, and that for the same price, we could fly around the world. Thus, we went by way of Europe with a stop in Paris to see the Galliens. We proceeded via Istanbul and Teheran before landing in New Delhi. We gave ourselves about a week of sightseeing with visits to Varanasi (Benares), Agra to see the Taj Mahal in full moonlight, and Jaipur. The meeting was in New Delhi where, in addition to seeing old endocrinology friends, we took in the local sights. Our return was by way of Tokyo for Lou's first visit to Japan and for me to consult with my friends at Keio University. It was here that we arranged for Jiro and Akiko Matsumoto to come to Tucson the following year. We enjoyed seeing John and Linda Taylor and their apartment in Akasaka Mitsuke. We had enough time to spend a day in Nikko to see in winter the beauty that I had experienced in June of 1961. We left Tokyo just in time for Christmas in Tucson.

There was much to digest after the long fascinating trip, and perhaps it was some distraction from the cancer incident of not many months earlier. Nevertheless, there was a wary watch kept to be certain that no latent cancer cells could ever start growing. With Lou's complete thyroidectomy, she needed a daily dose of exogenous thyroxine, and this was supplied at a level sufficiently high to suppress her own TSH production. Monitoring for any possible thyroid hormone production of her own was a painful proposition that she was required to experience from time to time over many years. As time passed she had fewer tests per year, but it was about twenty years before it was deemed that the tests were no longer needed.

At this point in my scientific career, there was enough progress to warrant a fair amount of international travel. In fact, during the next year (1968), I participated in two international conferences in Paris that were concerned with the two major aspects of my work, endocrinology and developmental biology. Accordingly, we spent two months in Paris, profit-

The Advisory Council of the International Society for Comparative Endocrinology at the Vth International Symposium in New Delhi, India, in 1967. Included are dear friends: top row, third from left, Howard Bern, and to his right, Aubrey Gorbman (both deceased); bottom row, third from left, Louis Gallien (deceased), second from right, Emil Witschi (deceased), while the still-surviving Hideshi Kobayashi is at the far right.

ing from the profound hospitality of the Gallien family. Big Lou gave me free rein of his laboratory even while he was mostly away. I was very much at home there. At the same time, Suzanne Gallien offered us her apartment near Port D'Orleans. I did help Big Lou in a minor way during the second conference, the VIth International Congress of Embryology, of which he was president. He asked me to serve on the local committee primarily by welcoming guests. It was no hardship to spend time at the famous Hotel Lutetia providing information and guidance to arriving Americans and other participants. We returned to Tucson late in the summer and, despite an urgency to be there for the arrival of the Matsumoto family, we felt it more important to stop in Iowa *en route*. Lou's father was not well, and we both sensed that he would not survive until we had another opportunity to visit Lou's family. Our premonition it seems was well founded, because Lou was called home in October to assist as her dad passed away.

9

The Vicarious Enjoyment of Field Biology

The more than ten years of scientific activity that had transpired to this point were mostly spent in the laboratory. However, given my personal interest in field biology and the presence of prominent vertebrate zoologists among our faculty, I took advantage early on of their knowledge and expertise. Naturally, my interest and need for amphibians led me to seek out our herpetologist, the late Charles Herbert (Chuck) Lowe. He was an enormous physical presence which, at times, he used as a means of intimidation. I soon learned that this was his way, so I paid little attention to such posturing. Chuck was extremely knowledgeable not only about herps, but about ecology in general. When I arrived, he was interested in physiological ecology, and he hoped that I would work in that area. I had my own research interests and declined to do so, although I would go out with him from time to time on one of his forays. One evening was particularly funny. We were driving along and spotted a toad hopping in the road. We got out, and Chuck grabbed the toad and immediately jammed a long thermometer into its cloaca. What a sight it was to see this huge man sitting in the middle of the road, legs splayed, and between them, this thermometer-impaled toad.

Although we did a few things socially with Chuck and his wife, Arlene, he and I never really connected well. I suspect that he resented the fact that I didn't join him in his interests in physiological ecology; moreover, the fact that I was working with frogs didn't set well. I am grateful to Chuck for his having bestowed a title on my little group. He referred to us as the "tadpole patrol." I know full well that he did not mean it as a compliment, but I was pleased with the title and used it from time to time. Some time later, one of my undergraduate premed students designed a tadpole patrol insignia. It adorns the door to my office to this day.

A bird story

Also, during these early years, I spent a fair amount of time in the field enjoying vicariously the research projects of other colleagues. One of these scientists was Joe T. Marshall, the outstanding ornithologist of our Zoology Department in 1956 when we arrived. If there ever was a person who truly deserved to be recognized as a character, it was he. There are so many people that we meet who have the reputation of being a character only because they work at calling attention to themselves. Joe Marshall was a natural, as I point out in a few vignettes. Joe received his training at the University of California in Berkeley. He had been a gymnast when he unfortunately contracted polio. He did not let the disease get him down as he continued his gymnastics. When we arrived, he was working out regularly at historic Bear Down Gymnasium. Joe dragged one leg when he walked, but that did not keep him out of the field. One of the projects he was working on with the support of an NSF grant was the pairing behavior of towhees. His study site was at Indian Dam along the banks of the Santa Cruz River on the San Xavier Indian reservation, quite near the renowned Mission San Xavier del Bac. In those days, the Santa Cruz valley was quite moist and supported a substantial mesquite bosque in addition to good stands of cottonwood trees. There were two species of towhees that bred there, the brown towhee, now called canyon towhee, and Abert's towhee. Joe would go out early in the morning to set mist nets to catch and band towhees. The characteristic bands allowed him to identify the birds on subsequent visits. Often, on Thursday mornings he asked me to go along to help with the mist nets. His work showed clearly that towhees pair for life, but that sometimes there were divorces and reconciliations.

I talked with Joe quite a lot and learned that he had been born during WWI when his father had been attaché to Gen. George Pershing in Paris. I never learned much about his upbringing or how he got to Berkeley, but I did learn much about his work. A project that he did before I knew him led to a monograph on the birds of the Rincon Mountains. I know the trails to Mica Mountain and Rincon Peak fairly well and fully realize what an effort it must have taken to do his study. Another project he worked on concerned screech owls, and after I told him that we had a family near our house, he came up to check on them. Indeed, we found the mother feeding young between our house and Bond's. I should point out that Joe had a musical ear and perfect pitch. He told a funny story about that in connection to his screech owl study. Joe and Elsie did not have a home of their own. Rather, they often rented the home of a University of Arizona faculty member who was on sabbatical leave. Near one of the houses they had stayed in, Joe heard the call of a local owl, and by mimicking its call, he got the owl to approach so close that he was able to reach out and grab it. Joe wanted a photo of the owl to use in the monograph he was writing so, soon after the previous incident, he went in search of the owl with camera in hand. He got the owl to return to his call and to approach closely, but never close enough for Joe to catch it again.

Joe's musical and other skills became quite well known in Tucson. I believe that it was someone in the then College of Fine Arts who showed Joe an antique Italian harpsichord that they had acquired and that needed total restoration. Joe took on the task. Harpsichords of that age were made from Italian cypress, and so Joe tracked down some of these trees in Tucson. Randolph Park was home to quite a few of Italian cypress that I always knew as "cemetery trees." He was able to work out an arrangement with the city to give him those trees that they were going to remove. Thus, he was able to obtain authentic Italian cypress logs from which to make useable wood. As part of the restoration, he needed authentic European boar's bristles, but this was no problem since the wildlife unit had such a boar's head mounted on the wall of their teaching lab. Joe was able to complete the restoration and to produce a beautiful sounding instrument. He was so successful that he went on to build other harpsichords from scratch that the College of Fine Arts was able to put to use. There are only a few people left who would remember this, but for quite some time one of Joe's harpsichords was on display in a window of Cele Peterson's dress shop, at that time located downtown on Pennington Street.

There were always stories of Joe Marshall going around. One of the more interesting ones is a bit mysterious, and of which there were several versions. It related to an event that must have taken place in 1960 when we were away in Europe. The way I understood it was that Joe, who always had an interest in behavior, came across a freshly killed Harris ground squirrel on the road as he was going to the university. It appeared that another squirrel was attempting to mount it. He took it in to the collections area and placed the dead female in the lordosis position in a cage. He was asking the question, does the assumption of this mating stance induce males of the species to mate? It seems that when he placed a male or males in with the dead female, they immediately tried to copulate with her. Enough data were collected to warrant the writing of a brief note for the *Journal of Mammalogy*. One of Joe's other interests and talents dealt with limericks; in fact, he was a veritable walking library of limericks, capable of reciting them from memory with ease. Many of us knew this and shared limericks with him. My understanding was that the editor of the *J. Mammalogy* was also a limerick virtuoso, and this prompted Joe to play a joke. The brief note that was submitted was entitled "Davian Behavior in Ground Squirrels." When the note was received by the journal, the limerick aficionado was not there, and the person who received it did not question the title. If he had, he would have realized "Davian" referred to a ribald limerick, "There once was a hermit named Dave who kept a dead whore in a cave. You have to admit, he hadn't much wit, but think of the money he saved." The brief note was published with this title and apparently caused much embarrassment. Over the years, this story was told with many versions. I had never seen the published note, but one of my colleagues, the late Bob Chiasson, showed me a reprint of it, and the surprising thing was that the author was Robert Dickerman. A year or so ago, Lou and I checked into the possibility of contacting Robert Dickerman, but have had no luck. Joe Marshall seems to be deceased, and there is no one else that we know of who could clarify what happened back in 1960. We did find that Robert Dickerman had been graduate student at the University of Arizona at the time; moreover, he was an ornithologist. It seems likely that he was a student of Joe Marshall's. An interesting outcome of all of this is that Davian behavior has become the accepted scientific term for the act of necrophilia where males attempt to copulate with dead females. It applies to animals of all classes.

A manifestation of Joe's true and profound interest in his science was clearly demonstrated by his suddenly resigning his position as a tenured full professor at the University of Arizona to pursue new ground. He felt that he had fully exhausted his ornithological calling in Arizona, so he moved to the Smithsonian Institution to explore pioneering work on mammalian behavior in Indonesia. It did not take him long to register success in this area when he published a fine paper in *Science* on orangutan behavior. Over the years, I have lost touch with Joe, but often after his departure from Tucson, he would return to visit. I saw him at least once, but at another time he stopped by my office. I knew that he had been there since he left a unique calling card. Joe and I frequently used to greet one another with the words "Jo-Jo the dog-faced boy." During the noon hour I was often away from my office to play handball, but I always left the office door open. One day when I returned, I found a small scrap of paper on the corner of my desk on which was written "Jo-Jo the dog-faced boy."

The bat cave of Eagle Creek

In addition to Joe Marshall, the ornithologist, and Chuck Lowe, the herpetologist, the third vertebrate zoologist of our department in 1956, was E. Lendell Cockrum, our mammalogist. Len had an active cadre of graduate students who were all busy with their own special mammal. A target of investigation by the Cockrum group was the array of bats that abound in southern Arizona. To that end, Len was funded by NSF to study the patterns of movement of perhaps our most common species, the Mexican free-tailed bat, *Tadarida braziliensis*. This species winters in Mexico, but summers in Arizona where the young are born. I had zero knowledge of bats when I arrived, but it took only a few months to learn something about them.

The Cockrums, Len and Irma, became friends, and we interacted with them socially. It didn't take long for me to learn what Len and his students were doing, and soon Lou and I accompanied Len and Irma on short bat outings. I remember the four of us going out, just after dark, down Mission Road past the San Xavier Mission to Helmet Peak, where Len would check on which species were roosting in a mine tunnel there. These tunnels are long since gone, as is Helmet Peak which was bulldozed away as the open pit mines took over this land.

Helmet Peak was small potatoes as far as bat colonies go, especially in comparison to some of the other bat caves. The granddaddy of such caves in southern Arizona is one located on Eagle Creek about five miles from the town of Morenci. This enormous cavern is home to an all-female colony of Mexican free-tailed bats. Moreover, it is a maternity colony to which pregnant bats from Mexico gather to deliver their young and spend the summer season. Eagle Creek originates in the southern part of the White Mountains of eastern Arizona and, along with other smaller streams, flows into the Gila River. The Gila and all its tributaries form an extensive riparian area that produces prodigious amounts of insects. The Mexican free-tailed bats are in their glory as they enjoy all this food.

Len's project involved banding as many as possible of these female bats with the hope that sometime later they would be found elsewhere. Thus, information about these banded bats, together with data derived from the recapture of bats that were banded in other locations, would begin to describe the annual movements of this bat species. Indeed, the banding and recovery project offered the possibility of obtaining a variety of other knowledge as well. The banding project offered many challenges. First, one had to get to the bat cave, and then the bats needed to be captured and banded. For all of this Len had to assemble a team and then transport the team and all the supplies to the bat cave. For the former, there were the Cockrum graduate students, together with other interested persons such as myself. Bill McCauley, one of our colleagues, was also an interested participant. Thus, we had three faculty members and several graduate students and friends. Another important member of the team was Bill Musgrove, a high school biology teacher from Kingman. Len rented a surplus Army gun carrier to carry our supplies over the rugged route to the cave. This meant driving through a streambed for about five miles from Morenci to the cave. The team members negotiated this journey as best they could.

The site of the cave was up a short incline from our campsite. As I recall, it was a warm part of the year. More than likely, it was June and thus before the summer rains. The cave entrance is rectangular in shape, probably about twenty feet high and ten feet across. The plan was to capture the bats at dusk when they were leaving for their evening feeding. In order to do this, a rectangular frame had been fabricated from aluminum I-bars. The frame was strung with piano wire in the long direction. This frame was suspended in the rectangular opening such that emerging bats would fly into the piano wires which they would not detect until the last minute.

After a collision, the bat would slide down the wire and fall into a wide funnel and into a short chute that fed a bushel basket that could hold lots of bats. Around each basket, one or two of us would be situated with a supply of bands strung on a rod. We all wore gloves and had been previously inoculated against rabies. The trick was to quickly reach into the basket, grab a bat and spread its wings with one hand, and clamp a band on the front wing edge with the other. The bat was then tossed in the air to its freedom. The whole exercise took only a few seconds.

Behind this facile description there were several problems. Perhaps the most serious was, how can the piano-wired frame be suspended? This was Bill Musgrove's task. We had brought with us a long 4x4 beam cut to the width of the cave entrance. There was a pulley attached to the center of the beam. Bill scaled the wall of rock alongside the opening, and when he found an appropriate point that was sufficiently high, we boosted up the beam. He then wedged it securely across the opening. One end of the rope was fastened at the center of one end of the frame, and the other was fed through the pulley. We hoped that we now had a means to haul up the frame into the mouth of the cave and to hold it securely in place during the bat banding operation. Luck was with us, for it worked just fine, and we were able to band and release lots of bats.

The next day was fascinating as we went into the cave and walked knee deep in guano. Someone had the mineral rights to the cave, but he had obviously not harvested guano for several years. The worst part of being in the cave was the constant mist of bat urine dripping from the ceiling. That year was apparently a banner year for bats. They completely covered the ceiling and the sides of the cave. We made some simple measurements and also counted the number of bats per square yard. When all calculations were done, the count came to 9 million! These were only females, and each had two infants, each attached to a teat.

We tried to clean up before the evening bat banding session. About an hour before sunset, we began to see raptors circling in the sky above us. There were red-tailed hawks, zone-tailed hawks, black hawks, and probably gray hawks and falcons. At the same time, within the cave, bats began to fly in full circles as though they were warming up. At a given moment as dusk approached a stream of bats emerged from the cave. It was time for us to set the piano-wire trap. In the sky, the raptors dove through the stream of emerging bats for an evening meal. The bats kept streaming from the cave for nearly an hour, but then, there were a lot of bats to exit the cave.

I participated in the Eagle Creek operation for at least two years and probably a third. It is hard to remember, because that was more than fifty years ago. However, the details I have presented are vivid in my mind as though they happened only yesterday. Unfortunately, bat numbers diminished greatly after these early years, in large measure because of the effects of DDT on insect populations.

Beyond a snail's pace

In the months before our departure for my sabbatical year in Paris, I often conversed with an elderly gentleman who occupied a room in the animal facility annexed to our building. It was here that I kept my breeding colonies of amphibians. I knew him as "Uncle Joe," but he was in reality Joseph Bequaert, a friend of Al Mead, our department head. Al Mead's field was malacology and Uncle Joe was extremely knowledgeable in matters malacological although he was primarily an entomologist. In fact, Uncle Joe had been curator of insects at the Harvard museum. When he retired, Al Mead offered him space in our animal quarters. Uncle Joe was a Belgian and a respected field biologist who had traveled extensively, especially in what was then known as the Belgian Congo. He was very nice to me and expressed interest in my forthcoming sojourn in Paris. He helped me translate documents, since my knowledge of French was pretty meager at the time. When we returned from France for the fall semester of 1964, I went to see Uncle Joe to impress him with my newly acquired ability to speak French. Of course, he was nice enough to act duly impressed. Furthermore, he told me that I had a French speaker in my embryology course. The semester had just begun, but I had noticed that there was an older man sitting in the front row. He was Walter B. Miller, the person that Uncle Joe had referred to. Walt Miller had retired, as Captain, from the US Navy and had started graduate school as a PhD student of Al Mead. During his childhood years, Walt had been fascinated by snails, and as an adult he became an accomplished and published malacologist. Al Mead had met him at a malacological conference and encouraged him to come to the University of Arizona for graduate work.

Walt was far and away the best student in my class (and in all the other classes he took). As I got to know him, it was obvious that his superb performance was in part due to a superior intelligence. Walt had graduated near the top of his class at the US Naval Academy. Moreover, his motivation to excel as a student was exceedingly great. There would not be many people his age who would be willing to forego the financial security that he had in order to become a graduate student. When we met Walt, he must have been about forty-six years old. At the time of his retirement from the Navy, he was commander of the Pacific Missile Range at Point Mugu. Although he could have stayed in the Navy to await a promotion to the rank of Admiral, the lure of academia won out. But, he needed to provide for his wife and two sons. Betty Sue was a homemaker, his elder son, Walter, was in college, and a younger son, Nick, was in high school. In order to save for graduate school, Walt took a position with IT&T for two years. I believe that when he resigned he was a vice president. It must have been a financial jolt to leave such a well paying job to become a low-paid graduate student.

In short order, Lou and I and Betty Sue and Walt became friends. While the four of us interacted socially, Walt and I connected scientifically. I knew very little about snails, especially land snails, but I soon learned a lot from Walt and Uncle Joe. I had no idea that, despite its arid environment, Arizona is home to many species of land snails, many of which were of some size. They survive by crawling into moist crevices or rock slides where they seal themselves to flat surfaces until favorable conditions arise. I was intrigued enough to accompany Walt on numerous field trips. Despite his advanced years, Uncle Joe sometimes went along. We had become fast friends and conversation never lagged. Walt had chosen to do a revision of the genus *Sonorella*, a moderate-sized land snail that inhabited much of southern Arizona. It was an appropriate subject for a PhD dissertation since most of the early work was done in the early 1900s and many of the species were named on the basis of shells alone. I had lots of fun helping Walt. Aside from driving and hiking to remote places, it was interesting to dig down into rockslides to find our target snails. We talked of many things and I learned the fascinating story of Walt's origin.

Walt was born in Paris in 1918. His father, an American named Miller, was part of a young group that volunteered to fight in WWI with France before the United States entered the war in 1917. My understanding is that this group, the "*Frappe de Lafayette*," was essentially reciprocating for what

General Lafayette did for us in the Revolutionary War. During his stay in Paris, the young Miller encountered a beautiful girl on Montmartre. The young Madeleine was only 15 years old, but they married, and Miller went off to the front. Madeleine had become pregnant, and was five months into her pregnancy when they learned that Miller had been killed.

Apparently Madeleine was eligible for benefits from what was the equivalent in those days to our modern Post Exchange. Since she was a beautiful young widow, she was well known to many of the young soldiers. After his return, one of the young men who had known her must have had amorous intentions toward her. So when one of his friends, a school superintendent named Nixon, indicated that he was going to France, the smitten one asked him to look up Madeleine and to extend a proposal of marriage. When Nixon found Madeleine, he himself was so taken by the young widow that he proposed instead. Thus, in a true Miles Standish-John Alden parallel, Nixon married Madeleine and took her back to Charleston, South Carolina. Madeleine left her little boy, Walter B. Miller, with her mother, and thus young Walt Miller was raised by his grandmother on Montmartre. When Walt was thirteen years old, Madeleine brought him to America. Of course, Walt grew up like any French boy and experienced the advanced educational system of the French. Since he was the son of a deceased American soldier, he was eligible to attend a military academy. He was still too young to enroll in the Naval Academy so after attending the Citadel, he matriculated at Annapolis. It seems that his life with Madeleine and his stepfather, Nixon, was agreeable. These were years at the height of the Great Depression during which time the various programs of President Franklin D. Roosevelt were in full swing. During the summer, Nixon was in charge of a Civilian Conservation Corps (CCC) project in the mountains. During this period, Walt was taken under the wing of a Forest Ranger from whom he learned about nature and especially taxonomic botany. This was easy for Walt for he had a natural propensity for nature. It was already manifested when he was a boy in Paris when he collected snails and raised them in a shoe box. Despite his interest in natural history, Walt majored in engineering at the Naval Academy. Nevertheless, after his graduation and throughout his service in WWII, Walt maintained his interest in snails at a professional level.

While I accompanied Walt on numerous snail forays, two stand out.

One concerned the search for a Sonorella species known only from a shell held at the Philadelphia Academy of Sciences. Some time in the early 1900s, this snail shell was sent to the Academy by an itinerant preacher who served various mining camps in southern Arizona. This preacher was an amateur naturalist who sent unusual specimens that he found to the Philadelphia Academy. He found this intact snail shell in the small mining town known as Richinbar. The shell was obviously from a species of Sonorella, but which one? Richinbar was located on the banks of the Agua Fria River, a bit north of Phoenix. It was not indicated on any of the current topographic maps, but Uncle Joe came to the rescue. He went to the university library and began to probe antique topo maps. He found one that showed Richinbar's location to be about a mile east of highway I-17 on the west bank of the Agua Fria river. Walt organized an overnight trip in which we would first find Richinbar and later spend the night in Cottonwood. This would put us in position to explore Mingus Mountain, near Jerome, in search of another lost species. Accordingly, one morning in early summer, Walt, Uncle Joe, and I drove north from Phoenix on I-17, past New River to a point that seemed to be directly west of Holy Joe Mountain on the other side of the Agua Fria. We found a locked gate to an old ranch road going east and followed it on foot to the cliffs of the Agua Fria. By good luck, we hit our target perfectly, for here we found the remains of adobe walls, obviously miners' homes, complete with bearded iris in bloom. This was what was left of Richinbar. In short order we discovered snail shells, and after turning over boulders, we were rewarded by the discovery of live snails.

Joe, the naturalist, on a snail foray in the Rincon Mountains.

Walt's dissection of some of the specimens showed later that the mysterious snail was actually the known species *Sonorella sabinoensis*. Bravo, for Uncle Joe for finding the old topo map and for his enthusiasm toward the venture. What was also remarkable is that, at his age, probably in the eighties at the time, he was able to make the trek to the Agua Fria. The second part of the trip to Mingus Mountain was not successful, but we returned to Tucson content with the Richinbar accomplishment.

My contributions to Walt's malacological forays did not go unrewarded. In the spring of 1967, Walt asked me to accompany him on a two-night trip to collect specimens of a Sonorella that he and his son, Nick, had discovered on Rincon Peak the previous fall. He had collected only a few specimens, but upon their dissection he discovered that it was a new species. He had written a species description and now needed additional specimens to provide shells for museum use. I did not need an incentive to make the trip, but the fact that Walt named the new species after me certainly was additional motivation. I was happy and proud to hike Rincon Peak in search of *Sonorella bagnarae*. I should point out that the manuscript he had submitted described another new species, *Sonorella bequaerti*, that Walt discovered on Tanque Verde ridge. That made me feel honored even more to be recognized along with the noted Joseph Bequaert. Thus, the trip was planned.

We were to leave in the early afternoon and climb to Happy Valley Saddle, a site between Mica Mountain and Rincon Peak. The shortest and most direct route was by the Rincon Trail whose trailhead was on the X-9 Ranch. From that point to Happy Valley Saddle was only a short five miles. Unfortunately, the owners of the X-9 had closed all access to trailheads on their property due to some unfortunate incidents, mostly caused by hunters. Fortunately, one of my undergraduate premedical students, Art Jansson, helped us with this issue. During previous summers, Art had worked for the Park Service in the Rincon Mountains and had come to know and earn the respect of the X-9 people. He made an appropriate contact and was told where they would stash a key to a gate that would allow us to drive right to the trailhead where we could leave Walt's car. On the appointed Friday, we drove in to the gate, Art found the key, and off the three of us went. As we passed through the gate, I noticed that there was also a walk-through gate that was closed and had no lock. There were a lot of cattle accumulated near the gate, but we paid little attention to them.

We made it to the saddle in short order and set up camp. We had a fine evening, and early the next morning we left camp for the peak following a short trail of just over three miles. The peak appears to be a large flat rock that from the valley seems almost like a short thimble. In the middle of this surface is a cairn marking the summit at 8,482 feet. Of course, we signed the visitor's book. Nearby was an enormous claret cup mammalaria cactus typical in sunny places of that elevation. We headed for a rockslide on the northeast side of the summit, and with a little digging, we soon found our sought-after specimens of *S. bagnarae*. The task didn't take long so we descended toward our camp. About halfway down, we encountered a man and his teenage son, and that there were other people on the trail, surprised us. We didn't engage them in conversation, but we assumed that they must have driven to the walk-through gate and entered in that way. We got back to our camp about noon, and at that point Art announced that he was going to continue down past the trailhead to the gate where his girlfriend, Robin, would pick him up. Walt and I walked about the saddle area enjoying a little natural history. After spending a second night at the camp, we descended to the car and home.

On Monday when I returned to the university, I encountered young Art. He said that when he returned to the trailhead on Saturday, "all hell had broken loose" near the entry gate. Apparently, the pair we had encountered on the mountain had left the walk-through gate open thus allowing cattle to escape and to wander all over. These must have been the cattle we had seen as we passed through. The X-9 cowboys had driven them there to be loaded on a stock truck to arrive on Saturday morning. The extra work and trouble in rounding up the strays apparently infuriated the X-9 people, and thus this event was "the straw that broke the camel's back." All access to the Rincon trails through the X-9 ranch was denied for years thereafter.

10

A Second Sabbatical— Italy, Our Second Home

An Established Scientist

Toward the end of the 1960s, my scholarly activities changed considerably as writing became a major focus. This came about because I was invited to write several review articles; however, the biggest task was to work on a book. When I was able to relinquish my endocrinology teaching to Mac Hadley, I felt that the endocrinology face of my scholarly profile was now to disappear. However, about that time I was contacted by C. Donnell Turner, who asked that I co-author the fifth edition of his book *General Endocrinology*. This was quite a surprise. Don was retiring from his position at Duquesne University to live in Nagoya, Japan, with his life partner. He felt that he needed help with the revision of his book, and he hoped that I would be the person to do so. I did not know Don very well, but I certainly knew the book. In fact, Don had asked me to contribute some figures for his fourth edition a few years earlier. I did not use his text for my endocrinology course, only because the text written by my dear friends Aubrey Gorbman and Howard Bern was more appropriate to my leanings toward comparative endocrinology. Don's invitation put me in a bad spot. First of all, my research interests were more toward developmental and cell biology, and I was actually trying to shed some of my endocrinological connections. Moreover, I had just divested myself of endocrinology teaching. On the other hand, Aubrey and Howard had shown no sign of updating their own book that was already six years old, and endocrinology was in the throes of a rapidly expanding literature. I called Howard, and then Aubrey for help, and their advice was unequivocal, "Do it!" They had no intention of doing a second edition of their book, but they still wished for the comparative aspects of endocrinology to be dealt with. They felt that I would cover that base if I accepted Don Turner's invitation. I did so, and thus the first textbook of endocrinology aimed at an undergraduate and non-medical student audience would continue for at least a few more years.

Don Turner's first edition appeared in 1948, published by W. B. Saunders Company, one of the more important presses of the time. Although I never had a course in endocrinology, I remember seeing it and referring to it as a graduate student. The book was unique and gained wide usage. The earlier editions were translated into French, Spanish, and Polish. At the time that Don wrote the first edition, he was a professor at Northwestern University but apparently was forced to leave there in 1948. I presume that Northwestern University was not willing to tolerate the fact that Don's life partner was another man. His subsequent years were spent at several other universities in several different countries. Finally, in 1960, he settled at Duquesne where he and his partner, Dr. Hiroshi Asakawa, resided until his retirement and their move to Nagoya, where Dr. Asakawa's family resided. As I look back, I am still grated by the cruelty that he must have experienced in those years.

Despite the distance and the fact that we had to rely on the international postal system to communicate, the fifth edition went well and was published in 1971. A lot of the work fell on my shoulders since Philadelphia, the home of W. B. Saunders, was a lot closer to Tucson than it was to Nagoya. I worked hard on the project and fortunately was able to enlist the help of Lou from time to time. She recalls typing some of the text for me in Gallien's lab during the late summer of 1970. Typing on a French keyboard with different letter locations was a painful experience for her. The sixth edition, published in 1976, fell entirely to me. Don's health had deteriorated to the point that he could not help, and he died in Nagoya the same year that the sixth edition appeared. It, too, continued successfully and was further translated into Italian; rights for a Japanese edition, published in English, were purchased by a Japanese publisher.

As if I had nothing else to do, another major publishing chore was pushed my way. In late 1969 or early 1970, my old friend Jim Ebert asked me to become managing editor of the *American Zoologist*, the official journal of the American Society of Zoologists, now the Society for Integrative Biology. The function of the journal was to publish manuscripts based upon specific symposia held at the Society's annual meeting. This journal filled a unique niche because it provided a means of presenting almost all that was known about a given subject at one time, viz. parthogenesis, comparative immunology, the pineal, animal behavior. I was flattered that Jim would ask me, especially in retrospect to my not having gone to his lab as a post-doc after I finished my PhD. Actually, I knew that Jim had long since forgiven me. Again, I demonstrated that I was a glutton for punishment by accepting the editorship, even though I knew that filling the shoes of Howard Hamilton, the original managing editor, was going to be a chore.

Getting settled

At the time that I consented to become managing editor of the *American Zoologist*, I was well into plans for my second sabbatical leave. Although there was no particular research need that called me to Naples, as a classical zoologist I strongly wished for an experience at the renowned and historic "*Stazione Zoologica di Napoli*," the first major marine station. It was founded in 1872 by Anton Dohrn, a German zoologist who was looking for a seaside research venue in a more temperate region. Over the years, one famous researcher after another from northern Europe went to the *Stazione* to do research. With time, it became the place to go, and soon its library became a beacon that attracted many. Despite the fact that we had solid marine laboratories in the USA, even Americans were attracted to Italy. Lou and I each wanted an Italian experience, and besides we had friends in Naples. Again, I applied for and was granted a Fulbright award. In addition, our own National Science Foundation was accepting research proposals to fund a research table at the *Stazione*. A research table, or "*tavolo di studi*," provided for a research lab and all the facilities and supplies for an investigator for a given period of time. In my case, I applied for and received an NSF-funded table for nine months. Thus, we went to Italy in September 1970 and stayed until May 1971.

The Acquario (Stazione Zoologica di Napoli).

Just as with our previous sabbatical, the Fulbright award provided ocean passage and thus we booked the *France* from New York to Le Havre. We took the train to Paris and, after a short stay in Suzanne's apartment, continued on to Stuttgart to pick up a Volkswagen beetle that we had ordered earlier. We purchased it from Hahn Motor Company, a sales garage we knew about from our stay in Paris in 1968. I do not remember much about our drive south into Italy except that the countryside had the look of late summer that featured beige fields in place of the fresh greens of springtime. In our later years, we have spent a fair amount of time in a part of Tuscany just south of Siena, but always in the spring, during May or June, when the fields are bright green, dotted with red poppies. It was these very same fields that we traversed in 1970 when the greens had changed to browns and stacks of hay had replaced the poppies.

The administration of the *Stazione*, or the *"acquario"* as it is known to Neapolitans was quite well organized and prepared for foreign guests. English was the common language of convenience. As soon as it was established that I was to be guest investigator, I began to hear from the *Stazione* staff about housing and other matters. It was established that they would find a suitable apartment for us, so that was a relief. Thus, we did not pay a lot of attention to a letter we received from the Contessa Marilyn Gaetani d'Aragona describing an apartment that she and her husband were willing to rent us in Naples. Obviously, Marilyn was American since her handwriting was typical of a person educated in America. We kept her contact information just in case there was a snafu with the accommodation that the Aquarium had in mind for us.

We arrived in Naples in late morning and drove right into the *cortile* (courtyard) of the *Stazione*. We were a bit uneasy about leaving our car there while we checked in, even though the vehicle was locked. In those days, no one left unattended a car with baggage visible anywhere in Italy, and especially in Naples. We checked in quickly, and someone came down to show us the apartment that was destined for us. It belonged to Professor J. Z. Young, a famous and accomplished neurobiologist. Of course, I knew of him because of his pioneer work, some of which was on the pineal. The apartment was in Mergellina, just west of the *Stazione* on the Bay of Naples. The apartment may have had virtues, but whatever they were could not be seen through the mess and filth that confronted us. Obviously, J. Z. Young was a slob of the first order and must have possessed an arrogance to match if he was willing to offer to rent an apartment in that state. We declined the *Stazione*'s apartment choice for us and moved to plan B, the possibility of the Gaetani d'Aragona apartment.

At this point, an introduction to the Gaetanis is in order. At this time, Gabriel Gaetani del' Aquila d'Aragona was a Professor of Agricultural Economics. He was also a Count, which accounts for his impressive name. The family had an opulent apartment that occupied almost the entire top story of a large and ancient *palazzo* (building) in the old Spanish section of Naples. It should be noted that in the past southern Italy was occupied by the Spanish Bourbons. The Spanish aristocracy that lived in Naples inhabited a promontory, Pizzofalcone, that looked down on the Santa Lucia shore of the Bay of Naples. We do not know when the Gaetani family acquired their apartment, but it must have been many years ago judging by the 19th century furniture it contained. We presume that the apartment was just a city home, since the major land holdings, and probably the ancestral home, were to the south in the neighboring province of Basilicata. That home was in a small mountain town, Sasso di Castaldo. Gabriel's father must have been relatively young when he died, and Gabriel, as an only son, was raised by his mother and aunt. We gather that he must have lived a pampered existence. During this time, as part of his formal education, Gabriel came to the USA for a period of study at Cornell University in Ithaca, New York. At Cornell, as a good Catholic, he joined Newman Club where he met and befriended Marilyn Yannick, an undergraduate from Hornell, New York, not far from where I was born and raised. After Gabriel returned to Italy, he soon found himself alone with the passing of both his mother and aunt. Gabriel, who has a wonderful sense of humor, loves to tell this story on himself. With both his mother and aunt gone, Gabriel had no one to take care of him, so he called Marilyn in the USA and proposed marriage. She consented and went to Italy, and thus Marilyn Yannick from Hornell became the Contessa Marilyn Gaetani d'Aragona. Whatever the details, this was the state of affairs when we encountered Marilyn and Gabriel in 1970.

When Marilyn took on her new life in Italy, she must have hit the ground running. She soon learned to speak Italian and assumed her matriarchal duties. These included managing things, both in Naples and in Sasso. The latter must have been a real chore, since the estate was an active farm producing olives, wine, crops, etc. We gather that the initial culture shock was alleviated by her befriending a group of expatriates, some of whom had similarly married Neapolitan men. One of the Americans was Helene Small who appeared to be a sometime secretary to Marilyn. Thus, when we followed through with the rental of the Gaetani apartment, we dealt with Helene. Marilyn was in Sasso busy with harvest time activities, and Gabriel was occupied with his university activities when he was in Naples. He spent parts of the year in Basilicata where he was an elected member of the legislature of the province.

We met with Helene who showed us the apartment. It consisted of two large rooms, a small kitchen, and a bathroom. It was not ready, however, to be rented. It was being painted and some of its furniture was being restored. There were no dishes or kitchen utensils, but Helene indicated that we were to go to Standa, a department store much like Target in the USA, and buy what we needed to outfit the place. It all seemed rather intriguing, and we were not at all turned off by the apartment as we assessed its possibilities. The building had a locked *cortile* where we could park our car, and there was even a rickety old-fashioned elevator. We needed a place to stay until the apartment became livable, and Helene had this problem solved in a more than satisfactory way. It was the expatriates to the rescue. One of the American women, Rita Di Martino, was married to Renato Di Martino, a lawyer. Renato's older sister, Giovanna, was a single woman who lived alone in a beautiful apartment overlooking the Museo Pignatelli garden not far from the *Stazione*. She was willing to rent us a room until our apartment was ready. The total arrangement was fine with us and thus the question of our housing was resolved. All the negotiations were done through Helene and we did not see or meet Marilyn or Gabriel until well after we had agreed to the rental. We did not meet Marilyn until Christmas, although we did accept cocktails with Gabriel shortly after the agreement, but probably before we moved in. He seemed to like Scotch whiskey, so we sipped on Chivas Regal in one of the antique-laden parlors that was one of a linear string of rooms that comprise their apartment. After seeing the arrangement of rooms in Marilyn and Gabriel's apartment, we could see how the small apartment they were renting us came about. Their apartment was way too large for the two of them, so they just cut off two rooms at one end to make a rental unit. We liked Gabriel very much and had the first of many pleasant conversations with him.

We moved into Giovanna's with pleasure. The apartment was light and airy, and it was a joy to hear the birds in the garden below. There were blackbirds that we knew from France, blue tits, finches, European robins, and black-headed warblers. We made our own breakfasts and ate lunch and dinner out. Giovanna's apartment house was on Rione Sirignano just off of the Riviera di Chiaia, a major street that runs along the inner edge of the Villa Communale. The Villa Communale is a major park that extends from Piazza Vittoria practically to the American Consulate. The outer margin of Villa Communale borders on Via Caracciolo, a multi-laned thoroughfare that runs along the sea to connect Mergellina with Via Partenope and the port of Santa Lucia. This major route is separated from the sea, or actually the Bay of Naples, by a wall of boulders that provides good perches for many recreational fishermen, sunbathers, and lovers. In the center of Villa Communale is the *Acquario* or, for my purposes, the *Stazione*. It is a beautiful location surrounded by many bushes and trees, most notably live oaks. Giovanna's apartment was just a few hundred yards from the *Stazione*. The Gaetani apartment was a bit farther, but within easy walking distance. In fact, it was an interesting walk uphill, first to the Piazza dei Martiri and then farther up on the via Chiaia. Along the way, via Chiaia goes under Str. Monte di Dio before continuing on to the major piazzas and via Roma, the main drag. To arrive at Monte di Dio from via Chiaia, one must take an elevator, or climb the stairs. This is what we had to do to reach the Gaetani apartment, because their *palazzo* was situated on a *piazzetta*, little square, Piazza Santa Maria degli Angeli à Pizzofalcone, at which the elevator and stairs arrive. It is not as complicated as it sounds, although Lou's mother could never understand, when she visited us, how you could be on one street, take an elevator, and then be on another street. In any case, the distance to and from the *Stazione* was not very far. Altogether we were well situated.

The stay at Giovanna's was longer than we had anticipated, but it was advantageous in some ways. Giovanna was exceedingly nice, and for Lou it provided a situation much like we had with Mme. Fallex in Paris. Giovanna knew a little English, but not much. Lou did not yet speak Italian; however, the two of them got along pretty well in French. In this way, Giovanna got some practice in French, and Lou was exposed to good Italian. Giovanna was intelligent, cultured, and well educated. This was a fine environment for Lou while I was at the *Stazione* every day.

Getting to work

My situation at the *Stazione* was superb. The people who worked in administration were kind and had reserved a spacious office and lab for me. My rooms faced the bay and I had my own balcony with a beautiful view of the bay and Capri in the distance. Not far from shore was anchored the immense aircraft carrier, John F. Kennedy, a jewel of the VIth fleet based in the Mediterranean. Basic laboratory services were excellent, and I was soon at home.

The various personnel that I dealt with were a lot of fun and appreciated my efforts in speaking Italian which improved daily. The permanent staff member who headed up my section of the *Stazione* was Dr. Rainer Martin, a very nice young German from Ulm. He was an electron microscopist who was impressed with the work I had done on chromatophore ultrastructure. Early on I was asked to give the *Stazione's* weekly seminar in their historic seminar salon. In attendance were the various researchers of the *Stazione* as well as faculty of the University of Naples. Among the latter were my dear friend Giovanni Chieffi and his students, as well as Giuseppe Prota, a young organic chemist whom I had met at the VIIth International Pigment Cell Conference held in Seattle the previous year. With him was his wife, Giovanna Misuraca, a cell biologist at the University. During our stay in Naples, Lou and I developed a close friendship with the Protas, Peppe and Giovanna. During my seminar, I discussed the dermal chromatophore unit and the role of iridophores and melanophores in producing blue color.

A day or two later, Rainer engaged me in a discussion about blue coloration and told me that a common fish they caught in the bay had blue spots on its back. It was, in fact, one of the electric fish, *Torpedo ocellata*. He asked if its blue spots were the result of the same chromatophore associations I spoke of. I indicated that I did not know and that the question could only be answered by doing a study of *Torpedo* skin. We decided to do a little collaborative study with Rainer doing the histology and electron microscopy that might be needed. It eventually turned out to be an interesting project that demonstrated the blue color to be produced by means entirely different from those producing blues in frogs. Instead, these blue spots were very much like blue nevi found in human skin. This blue was a structural color.

The torpedo project was purely a sideline. The work I had planned was actually a throw back to my early work on the pineal. Much of the pioneering work on the pineal had been done on larval lampreys (ammocetes) which were plentiful in some of the rivers that emptied into the Mediterranean near Naples. I was interested in a pineal-pituitary association and some of the experiments would use hypophysectomized ammocetes. I don't believe that anyone had successfully performed this operation on these larvae that are about the size of a pencil, but I thought that I'd give it a try. An animal supplier named De Rosa and his son collected experimental animals for Chieffi and his colleagues at the university, so I contacted him. Off they went to the Volturno River, just north of Naples, and brought back a nice supply of *Lampetra planeri*, the species I was looking for. Hypophysectomy went surprisingly well, and I don't recall too many deaths from my efforts. It seems that I had a fair degree of success with the surgery, because most of the larvae I operated on became pale in a short time. Moreover, these pale animals injected with MSH immediately darkened to their original coloration.

Perhaps the most significant of the research efforts that stemmed from our time in Naples involved little hands-on activities from me. I refer to the start of several-years collaboration with Peppe and Giovanna. When I met Peppe the previous year, his presentation in Seattle dealt with the chemistry of phaeomelanin and related pigments that are found in red hair. As a student of the noted Neapolitan chemist, Professor R. Nicolaus, Peppe extracted and elucidated the chemical nature of the red pigments from the hair of a family of redhaired people that lived in one of the *vicoli* (tiny streets) near the university. The family had lots of redheaded kids and was happy to sell their hair, thus providing the Nicolaus lab with much material. I was fascinated by the story, primarily because it was interesting, but also because I had a red pigment problem of my own that needed solving. While we were working on the dermal chromatophore unit project, John Taylor and I were very much surprised to find that the melanosomes of the Mexican leaf frog were remarkably different from those of other species. They were much larger than others and were of a compound nature that featured a core of eumelanin surrounded by a concentric layer of dense fibers that were red in color. After John had returned from his fellowship in Japan and was appointed assistant professor in the biology department of Wayne State University, we published our observations on this melanosome. This was to be the first of many publications on these unusual melanosomes. While John was in Japan, I began work on the fibrous component of the leaf frog melanosome and discovered that the red color was due to a red pigment that was readily soluble in alkali. I knew that the pigment could not be phaeomelanin, but I hoped that Peppe, as a good organic chemist, would be interested in elucidating the structure of our red pigment. Indeed he was, and so I arranged for skin samples to be sent to me from my lab in Tucson, and the project began.

In addition to original research, editorial activities began to demand time as my role as managing editor of the *American Zoologist* provided manuscripts for me to review. Thank God, Lou, who is a better copy editor than I, pitched in to help. The work provided something to occupy some of her time, and it got her out of the apartment. One of the worst parts of the job was the mailing of manuscripts back and forth to the USA. In the beginning, I was quick to blame the Italian postal system for some of the problems we had, notably postage due. I soon recognized, several times to my embarrassment, that the Italian postal officials in Naples really knew the international postal regulations very well. It seemed that problems often arose because this was not the case with some local post offices in the USA. When we returned home and utilized the US postal system almost exclusively, we were much happier. My job got considerably easier after we returned home since Lou did more and more of the work, and we learned a little more about what we were doing. My job as editor was of course strictly voluntary, but the American Society of Zoologists paid Lou for her work.

Getting to know Naples

In addition to a certain amount of work on my part, our stay in Italy provided a lot of pleasure. In those days, Naples was still relatively poor and had a bad enough reputation to preclude the tourism that it deserved. That was unfortunate because the cultural, historical, and architectural wealth of the city and its surroundings are superb. Everyone knows about Ercolano (Herculaneum) and Pompei, but the richness of the areas just west of Naples was then untapped. Even today, these many treasures escape most visitors. Lou was devoted to her Michelin guide and planned many short visits for us. Our Volkswagen beetle was handy as we drove to Bacoli and Baia, home of the Roman fleet. From the guide we learned that the Romans had dug an enormous cavern in the tufa, a soft, volcanic ash-derived stone that underlies the area. Bricks of tufa were a fundamental building material that the Romans used. This cavern was, in effect, a gigantic cistern in which fresh water was stored to supply the fleet that was anchored at Miseno, which is at the very tip of the peninsula that defines the west side of the Bay of Naples. The cistern was appropriately named *Piscina Mirabilis* (wonderful pool). It is a beautiful structure that was supplied water by a very long aqueduct that descended from the hills and mountains above Naples. We were able to find the *piscina* by following the weird procedure described in the Michelin guide. In effect, we were guided down a small street and instructed to find a woman who had an ice cream stand. It was she who had the key to the gate of the *piscina*, and for 100 lire (about 15 cents in those days) she took us down a slope to the gate and let us in. What a magnificent sight it was. The cavern was empty of course, but with the help of rays of light radiating through cracks in the ceiling, we could see several large arches supported by tufa columns. Water dripped in places to feed vines that hung down from the cracks above. While the guidebook knew about *Piscina Mirabilis*,

Inside the ancient Roman cistern, Piscina Mirabilis, located in Bacoli where the fleet was anchored.

we were amazed that none of our Neapolitan friends had knowledge of it. One Sunday, we took the Prota family on a picnic in the area. We asked if they would like to go to the *Piscina Mirabilis*. Both Peppe and Giovanna were born and bred Neapolitans, but they had never heard of the place. Of course, just as we had been, they were immensely impressed with this beautiful piece of history. A few years ago, some forty years after our first visit, we took some friends, also Neapolitans, to see it. To our amazement, despite some improvements and the addition of some safety features, the place remains the same. One needs to be admitted by a woman, who I believe was also selling ice cream. When I told her about our first visit so long ago, she replied, "That was my mother" who still minds the key from time to time. She took us to see the old woman.

Across a little bay from Bacoli is Pozzuoli, a seaport of ancient origins founded by the Greeks in the sixth century BC. It later became important to the Romans who built impressive monuments that still survive. One of these, the Serapeum, apparently used as a market, was part of an ancient Greek agora. Over the centuries, the volcanic phenomenon known as bradyseism left its mark. Its effect, common to the area, is based upon a slow raising and sinking of the ground, such that the bases of buildings close to the sea fall below sea level and remain so for years. Thus, the columns of the Serapeum are covered with mollusc shells to the height of a meter or more. Modern Pozzuoli also has its fame, at least as far as I'm concerned, for it is the hometown of Sophia Loren. Around the corner on the Mediterranean side of the Capo Miseno and further to the northwest is Cuma, an even greater relic of the Greek occupiers. Cuma was founded in the eighth century BC and was the home of a famous Sibyl, one of the most venerated in antiquity. At its height, the Oracle of Cuma rivaled or exceeded the Oracle of Delphi. We have fond memories of our first visit to Cuma because we were accompanied by our dear friends, Howard and Estelle Bern, who were visiting in Naples at the time.

As winter approached, we still had not met Marilyn who remained in Sasso; however, we shared Scotch whiskey with Gabriel on several occasions. We did talk with Marilyn by telephone a few times. In fact, it was she who urged us to buy whatever household furnishings we needed. In one of her calls, Marilyn invited us to come to Sasso di Castaldo for Christmas. We gratefully accepted, so as Christmas Eve approached, off we went south to Basilicata. Things got pretty primitive as we left the Autostrade del Sole and drove into the more rugged interior. The Gaetani residence was easy to find, and we soon met Marilyn and Caterina, who was in charge of the domestic end of things. Much like their apartment in Naples, the country home was furnished mostly with antiques. At the start, only Lou, Marilyn, Caterina, and I were there because Gabriel was not to arrive until Christmas Day. He did not drive, but had a driver available at all times.

Christmas Eve dinner is, of course, exceedingly important and features a fish dish. For those who live near the sea, such as our Neapolitan friends, *capatoni* (eel) is the fish of choice. In fact, for days before we left for Sasso, there were vendors on the streets selling live eels from shallow basins. In more remote areas such as in Sasso, *baccala* (cod) was served. I do not remember what else we had, but certainly a first course of pasta with a meatless sauce, and a salad of some sort. What really stands out in my memory was the pecorino, a cheese made from sheep's milk. This cheese is cured and aged in coarse, round, covered baskets, and thus its surface bears the imprint of the woven reeds from which the baskets are made. When we were ready for the cheese course, Caterina brought out the cheese, no doubt from their own production, and began cutting pieces. After her first slice, I knew we were going to have fun, for I saw the first maggot. It did not deter my biting into the delicious cheese. Soon, several more maggots were crawling about the cheese board. When Marilyn saw them, she let out a shriek. It seemed that she was not yet accustomed to life in the country. The maggots were no problem for Lou and me after having lived in France. We well recall a similar incident in Paris when Lou's sister Dorothy was visiting us. We had gone to dinner in a rather nice nearby restaurant, and the three of us had requested one of our favorite cheeses, Munster, a soft ripened cheese. It of course also harbored maggots, and when Dorothy was almost through with her piece, she discovered a maggot or two on her plate. She was horrified and complained to the waiter who informed her, *"Mais, Madame, le fromage est bien fait comme ça."* (But Madame, the cheese is well done in this way.) I'm not sure whether either Marilyn, in Italy, or Dorothy, in France, were comforted by my explanation that these fly larvae were not dirty, but were, in fact, filled with good cheese.

Neither Lou nor I are church-going people, but the cultural opportunity to attend Christmas mass in Sasso was really appealing to us. Besides, Marilyn is a devout Catholic, and it seemed more than appropriate to accompany her to mass. What a memorable experience it was. The church was full of kids to whom reverence was not yet a part of their growing up. They dashed about even onto the altar as the priest said mass. I remember one little boy with a toy pistol in his hand running past the old priest who smiled kindly and patted him on the head. The whole service was altogether fascinating. Later, Caterina prepared a traditional Christmas Day *pranzo*. I do not remember what we ate, but I do recall being stuffed. We finished our meal just in time to witness that one of the relics of feudal times still existed in rural, southern Italy. Gabriel, as representative to the legislature, was to be paid homage on Christmas Day. After *pranzo*, a parade of locals came to the door to wish Conte Gaetani a happy Christmas. Of course, each one had a favor to ask *il Conte*. Usually, it was for the count's help in finding employment for a son or son-in-law. After dealing with a few supplicants, Gabriel announced that he was going to take a nap, and he left me the job of dealing with the constituents while he had a snooze!

On the day after Christmas, Lou and I took a stroll after breakfast. We had noticed that at one end of town there was a steep incline topped by the ruins of an old castle. It was the remains of a Norman castle, one of many that top hills for much of the length of the central mountains in the Italian peninsula. These castles were among the gifts that the Norman invaders from the north left in Italy as they moved south seeking new fiefdoms.

Among the other gifts that they left were their genes that took their place in the Italian gene pool. The presence of many red haired, blue eyed people as far south as Sicily, or in particular places such as Pozzuoli, bear witness to these genetic gifts. At any rate, as we strolled upward we were dismayed by the sounds of squealing pigs that echoed from farm buildings *en route*. We learned that this was the season for butchering pigs as stocks of salami, prosciutto, and sausages were being put up for the winter and coming year.

When we returned to the manor, it was decided that the four of us would drive to Potenza, the nearby capitol of Basilicata. Gabriel wanted to show us this small city where the legislature convened. Moreover, during the Christmas season Potenza was famous for its *presepe* (creche) composed of life-sized figures. It was housed in an enormous plastic bubble that looked like a giant mozzarella. During the season, visitors streamed in from the countryside to see it. The southern Italians and in particular the Neapolitans are noted for their production of figures used in creches. It is fascinating to stroll through the narrow streets of the old part of Naples, near the *spacca Napoli* to see artisans at work in street level shops that once were *bassi*, the homes of poor families. Many of these artisans were producing gorgeous little creche figures. Perhaps a more important reason for our trip to Potenza was a political one. Gabriel wished to extend holiday wishes to the Colombo women, sisters of the then Prime Minister of Italy, Emilio Colombo. We were happy to go along.

Our fascinating Italian Christmas was followed a few days later by an eventful New Year's Eve celebration. Giovanni and Annamaria Chieffi invited us for New Year's Eve, not at their home, but rather at the home of Annamaria's brother, Massimo Valentino. Annamaria's family, the Valentinos, were sweet, wonderful people. We had already met her sister Flora and probably Massimo as well. He was a dear man who owned a quality retail shop on via Chiaia, just down the street from the *ascensore* (elevator) that took us up and down to our palazzo on Santa Maria degli Angeli. At Massimo's, we met their mother and others of the Valentino clan who, together with the Chieffi family and us, made a large crowd. We ate, played games, and talked until midnight. It was interesting that everyone participated in card games and games of other sorts very much as I remembered from holiday gatherings when I was a boy in Rochester. This must be a traditional activity of southern Italians, be they from Naples, Sicily, or Calabria.

At midnight, all hell broke loose as a large din arose in the city as whistles, sirens, and fireworks sounded from every corner. Windows opened everywhere, and in true Italian tradition, debris was thrown out and onto the streets below. In the morning, crews were about early to gather up the trash, mostly old furniture! What a memorable New Year's Eve!

The serenity of New Year's Day morning was quite a contrast to the night before. We decided to take advantage of the tranquility and go up to the via Roma for a stroll in the absence of crowds and traffic. The via Roma, also known as via Toledo, a name remaining from Spanish times, was not far from our apartment. The via Roma is one of the main streets of Naples and certainly one of the main shopping streets. We loved to walk there and window shop. On this crispy cold morning, we paused to study the elegant shoes on display in a shop that cornered on the via Roma and one of the many narrow *vicoli* that led into the poor residential areas above. Suddenly, a Vespa scooter with two kids aboard brushed by, and the passenger snatched Lou's purse as they headed up the *vicolo*. I gave chase, but was no match for the Vespa that quickly disappeared into a rabbit warren of old apartment buildings. Lou was a victim of the common crime known as "*lo scippo*" that was so common in Naples and other Italian cities in those days. Fortunately, Lou had left her passport at home, so she lost only a little money. The main damage was to her psyche, but that was assuaged a bit by the police who were very kind to her. As Italy has become more prosperous over the years, this type of delinquency has diminished, but still exists and is especially bad for older women who cling to their purses to their own detriment. About ten years ago, our friend Marilyn was accosted and, in the process, fell and broke a hip. Fortunately, the joys of Naples far outweigh its negative side.

Among the joys for us were the friendships we made. We have particularly fond memories that stem from our early stay with Giovanna De Martino. Lou interacted with her more than I, since she and Giovanna spent time together during the day. On one or two of the Saturday evenings that we spent at Giovanna's, she invited us to be a part of a little social gathering that she had with old friends of similar intellectual interests. These included her widowed sister, Anna Marino, her brother Renato and his American wife Rita, Rosetta Panain, and Gigino whose last name I can't remember. The latter were probably schoolmates, and we think that Gigino was probably an old beau. After a lovely *cena*, we would just sit and talk. Since this was all in Italian, it was a bit of a chore for both of us, but it helped us learn the language better. Even after we moved into our apartment at the Gaetanis, Giovanna continued to have us over on Saturday nights, and we became members of this little social club. We stayed in contact with Giovanna for several years until her death, and we continued to correspond with Anna for a long time after that.

Not long after our return to Naples from our Christmas visit, Marilyn came back to Naples and we saw her often. The Gaetanis were kind and outgoing. It was not too long before we became friends as well as neighbors. We learned that Gabriel had just built a small villa on Capri and that they had been in the process of furnishing it. I am not certain of how legal their Capri project was; however, it probably never would have been challenged given Gabriel's social and financial status. Moreover, Gabriel knew how to get things done. The commune of Capri was very rigid about building on the island, and I believe that, at the time, there was a building moratorium. Capri is a fantastically beautiful place, but it has the severe problem of having no water of its own. Rainfall and water that could be brought from the mainland were all that they had over the centuries. A few years earlier, an underwater pipeline was installed across the Bay of Naples to supply Capri. Marilyn and Gabriel invited us to spend a weekend with them at the villa toward the end of the completion of the project. We accepted and had a fine time despite the fact that it was cold and clammy. Capri sits out there in the Bay of Naples, and it can be very cold and damp in the winter. This problem was compounded while we were there because the walls had just been plastered and were still wet. Subsequent visits during the spring and summer over the years have been heavenly by comparison. Whenever I was in Naples for teaching or for conferences in later years, keys were made available to me to use the Gaetani apartment or to stay on Capri even if the Gaetanis were away.

As springtime approached, we were able to persuade Lou's mother, Hazel, to come to visit us in Naples. Thus, at the age of seventy-six, she took her first trip overseas. This was quite an adventure in 1971 for a woman who was born and reared in rural Iowa. The Italian culture was a bit overwhelming at first, but she adapted pretty well. Soon she even began to like espresso—this was quite an accomplishment for a person who would normally drink coffee so weak that one could see and identify a coin at the bottom of the cup.

We took her everywhere in our little Beetle. I well remember an outing on the Amalfi drive. It was a beautiful sunny and clear day. In those days, the route was really narrow, and the curved and crooked road hung even more perilously over the sea below. She enjoyed it immensely. I believe the high point for her may have been the ascent to Ravello on the multi-switchbacked road that takes one through the Valley of the Dragon to the town. When we arrived at the piazzetta, there was more activity than usual for a weekday. Then we noticed that there was a funeral in progress at the main church that faces the piazzetta. It was a real treat for Hazel to see the immense floral wreaths propped against the wall of the church. One of the main features of Ravello is the garden of the Villa Cimbrone. It's a fairly substantial but beautiful walk to traverse their length that terminates in a flat, tiled vista point that overlooks the Gulf of Salerno. The view downward is spectacular, as we see the ribbon-like road of our ascent and the many terraced vineyards that produce the noted white wine of the area. This is a special place for us. At intervals along the wall of the overlook are ceramic statues of what appear to be female deities. I took a photo of Lou and her mother on either side of a statue, and we have ever since referred to it as the "Three Graces." We have returned to Ravello countless times over the past forty years, and have often repeated the Three Graces photo with friends and relatives including Lou's sister, Dorothy, our late niece, Sara, and friends Katie and Lucinda.

Hazel's visit coincided with Easter, so we took her to Rome to be there on Easter Sunday. Our good friend Flora Valentino, Annamaria Chieffi's sister, invited us to stay in her apartment in Rome, since she was going to spend Easter with her family in Naples. We drove to Rome and had *pranzo* with Flora before she left. We remember well the *pasta putanesca* that she prepared, and Lou has ever since followed her recipe. On Easter morning we made our way to Piazza San Pietro along with the hundreds of thousands of others that made up the throng witnessing Pope Paul VI. It is interesting that at several points at the periphery of the crowd, first aid stations had been set up. There was good reason for this, because we viewers, many elderly, were standing in the sun for almost three hours, all in close proximity to one another. Often, someone would faint and be carried to the nearest tent. Finally, it was Hazel's turn, and as she collapsed, a man next to her grasped her before she fell. We took her to a nearby tent, and within a few minutes she revived. It didn't take her long to recover, and within a half hour she was strong enough to down a pizza at a nearby pizzeria. After the stay in Rome, we drove on to Tuscany for a visit to Pisa and an overnight stay in San Gimingano. In those days, there were no crowds of tourists, so unlike today, we could drive right into town and to the hotel. After a night in Florence, I left Hazel and Lou there for a few days so that I could return to my work at the *Stazione*. The night before we were to take her to Rome for her flight home, Hazel said, "I don't want to go. I'm just getting acquainted. I know this [building] is a *palazzo,* and where you park the car is the *cortile!*"

Fulbright service and going home

Our nine-month stay was really a whirlwind of activity. My research projects were straightforward and did not require my total attention at all times. In part, the time was taken by the collaborative projects that I was engaged in. These included the torpedo project with Rainer Martin and the leaf frog red pigment effort with the Protas. An additional amount of time was devoted to Fulbright activities that I had not experienced in France in 1964. By 1970, Fulbright scholars were asked to extend their roles as visiting scholars. For at least those scholars in Europe, a list of their names and their interests was circulated among universities, and the Fulbright Foundation funded visitations to specific universities that wished a scholar to visit. Thus, I was invited to England to consult and to give lectures on the pineal in Durham, Salford, and Sheffield. These invitations were a real surprise, because I didn't know that the program existed, and I knew no one in either Durham or Salford. I did have comparative endocrinology colleagues in Sheffield, and it was they who invited me. Another surprise invitation came from Greece. Apparently, Marietta Isidorides, with whom I had not been in contact since our 1960 congress in Copenhagen, saw my name on the Fulbright list and thus initiated the invitation. She and I had a fabulous reunion in Athens. I was pleased to meet the head of her department, Professor Vassili Kiortsis, whose name I knew from the literature. His work on limb regeneration at the University of Geneva was extensive. I was surprised to meet him in Athens, since he had been such an established Professor in

Geneva. His explanation for the move to Greece I understood fully, having been exposed to Swiss culture from other sources. Kiortsis explained that he was Greek, and while this did not affect his academic status in Geneva, his children would be precluded from professional success. I believe that at least one of his children was studying law and the possibility for him to have many clients would not be great because the family was Greek, and not Swiss.

Speaking invitations outside of Italy were not all Fulbright based. The Galliens asked that we visit them in Paris, and a colleague from the University of Kiel, in Germany, asked that I come and give a seminar. In addition, I had an open invitation to speak at the University of Cologne. The latter invitation was rather interesting. It came from an investigator, Professor Walter Weber, whom I met one day at the *Stazione*. He had come down from Cologne for a brief stay. When he learned that we were planning to go by car to Paris and Kiel, he asked me to visit them and to stay in the guest room (*gast zimmer*) of the Zoology Department. Since Cologne was right on the way back to Naples, I accepted. So we drove from Naples to Paris, where I spoke at Big Lou's laboratory in French, and in Alfred Jost's lab in English. Freddy Jost spoke excellent English and was trying to encourage his disciples to follow suit (although in reality he probably didn't want to suffer my French).

It was in the season of Nouveau Beaujolais, and Claude had purchased a *carré*, a twelve-liter collapsible cube of the wine, from one of his friends from PUC. During our short visit, we drank the whole container, mostly at Suzanne Gallien's apartment during a soiree she hosted for Lou and me. There were seven of us, including Claude and Michèle and their close friends Michel and Nicole Berthelot, whom we knew from my days at PUC. I had to speak at Jost's the next day so I took it easy, but Michel Berthelot was absolutely *noir* (absurdly drunk). From Paris we headed to Kiel for a short visit and seminar. It was a quick trip down to Cologne from there.

We found Walter Weber and his lovely wife Ruth, and after a quick introduction—Lou had not met Walter and I had not met Ruth—Walter took us to the zoology building for a short rest before dinner. During our introductions, Walter indicated that he had spent time in Safford, Arizona. We speculated about this while we waited for his return, and given his age, a few years older than I, we concluded that he must have been a POW in Safford. We had a lovely dinner and met their two daughters. It took only a few minutes before we felt absolutely comfortable and at home. Walter and Ruth were gracious hosts, and right then and there we developed a friendship that lasted for years. In fact, Walter and Ruth visited us as houseguests in Tucson. During our evening, our suspicion was confirmed. Walter indeed spent several months in Safford as a POW, picking cotton. In WWII, Walter was conscripted into the Nazi youth movement. I believe that he was seventeen years old at the time. During his training he was told absolutely not to be taken prisoner by the Allies for they would treat him badly and that he would be starved. He was in fact captured by the British and instead was fed the first good meal in months. After a lovely evening with the Webers, it was time to go since the drive back to Naples would be a long one.

And so, our fascinating, eventful, and unforgettable nine months were drawing to a close and we needed to prepare for the voyage home. Again, Fulbright provided passage by sea and reservations were made for a sailing of the Michelangelo from the port of Naples to New York. This was a wonderful way to travel for it was possible to take our car with us as baggage. Thus, in New York we disembarked on one pier and our car was unloaded on an adjacent one. For shipment, the car could contain only a minimal amount of fuel, just enough to get to the first gas station, so after customs and immigration were through with us, we were on our way. In packing for our return, we had a few obstacles to surmount. First of all, it was made clear that we should not put baggage in the Volkswagen, because anything we left there would be stolen. Apparently it was a first order fact that even the tools and first aid kit that came with the car would be gone. We did take one chance with baggage that fortunately ended well.

During my stay at the *Stazione*, I was fascinated by the many Greek amphoras that were scattered about as decorations. These were genuine archaeological items that came from the sea bottom in the bay. Among the staff members of the *Stazione* were *sommozzatori* (divers) whose job it was to collect specimens for the various investigators. In doing their work on the sea bottom, they frequently came upon all sorts of ancient relics. They commonly found and collected amphoras or even Roman anchors. The supply was sometimes renewed with the uplifting of the sea bottom during periods of bradyseism. These ancient mollusc-encrusted relics are fascinating and attractive. I became friendly with one of the *sommozzatori* (a little knowledge of Neapolitan dialect helped), and I asked if he could find me an amphora as a souvenir. He reported back that this was not a good year. My disappointment must have registered, because a few weeks later he presented me with a well-encrusted Roman anchor. It was probably from a smaller vessel judging by its shape and size. Anchors were cut from tufa and were rectangular or triangular. The former were for larger ships. In the center at one end was a hole through which a line was tied. At the opposite side of the rectangle were two holes through which lead rods were embedded. When they were found, the lead rods were still in place having survived the centuries under the sea. The anchor that I received had holes in the proper place, but the wooden rods that filled them had long since deteriorated. So, we had in our possession this heavy piece of stone. How were we to get it home with our baggage? We decided we would leave it in the trunk of the car with the hope that potential looters would not be interested in taking this heavy piece of stone. When we arrived in New York, there it was in the trunk, still wrapped in dirty gym clothes in a shabby old briefcase. My *sommozzatore* friend showed me another kindness before we left Naples. One day, he presented me with a small newspaper-wrapped package. In it was a beautiful, small terra cotta vase. It had been given to him by a farmer who had discovered it in a small Roman tomb he found while plowing on his farm in the Roman zone along the Domiziana near Capua, northwest of Naples. Farmers often found Roman or Etruscan artifacts, but were reluctant to report their discoveries, for fear of having their land taken over by the government. The beautiful vase is a prize possession, and of course it came home in our hand baggage.

 The day of departure came to pass and we made our way to the pier, not far from our apartment on Piazza Santa Maria degli Angeli. Several of our friends were there to see us off, making our departure bittersweet. As the ship headed northwest, we said goodbye to Vesuvius behind us and to the many monuments we had come to love. There was Villa Communale with the *Stazione* to the starboard and Capri on the portside sitting splendidly in the bay. Our destination was Genoa, where we would pick up more passengers and, no doubt, more crew. In those days, more than forty years ago, the crew was mostly Europeans, unlike today when Southeast Asians and Philippinos in large numbers comprise the crew. On this Italian ship, the crew was made up of mostly Neapolitans and Genovese. The passage to New York was a lot of fun, in part because we had such a great interaction with the crew. In fact, at one point one of our fellow passengers asked, "Who are you?" We pointed out that these friendly crew members from Naples were absolutely amazed to hear Lou, a very American-looking woman, speak Italian to them with a heavy Neapolitan accent! All the Italian language that she knew was what she learned in Naples.

The Statue of Liberty was a welcome sight as we entered New York Harbor. Soon, we retrieved our Beetle, gassed up, and headed up the Hudson *en route* to Rochester for a brief visit with family. We had not seen John and Linda Taylor in their new home, so we headed across southern Ontario to Detroit for another visit and to see how John was situated at Wayne State University. From there, we drove south to Arlington, Tennessee, to stay with our Tucson neighbors, the Bonds, and to pick up our new Ford that we had ordered from the Bond Motor Co. I had hoped that Mr. Bond would have given us a superior deal, but I'm not sure that it was that good! We had a lovely visit at their home and experienced real southern hospitality and cuisine. We crossed the Mississippi and headed west in our new cars. We had sold both of our old cars when we left for Italy. The trip through the Bible belt was quite a change from what we were used to, and if NPR existed in those days, their stations were certainly not accessible to us in Arkansas or Texas. We had a bit of a reprieve from the boredom of this segment of our return when we stopped for a visit with Elena (Brown) in Arlington, Texas. It had been a little over ten years since Bill Brown had passed away, and we dearly wanted to see Elena. She looked great and was happy to be back in Texas in the company of her son Jack and his family who lived next door. At last, we got to see her famous Mexican restaurant, La Tapitia. Two days later, we drove in to Tucson and were able to say, "Home at last."

11

The Epoch of the Mexican Leaf Frog

The years following my return from Naples were surely the most active of my professional life as my research efforts touched upon a variety of themes and my activities were many. An overlying program throughout these years that extended almost to my retirement dealt with the Mexican leaf frog. Little did I know in the late 1960s that this beautiful green frog would offer so many stimulating research venues for me, my students, and associates. Perhaps what really paved the way for this involvement was the return of Lee Stackhouse. Following his post-doctoral stint, Lee obtained a position as Assistant Professor at the University of Toronto. Unfortunately, he was like a duck out of water in this august and tradition-bound setting, so unlike his roots in southern Indiana. He longed to return to Tucson. When he did, his latent entrepreneurial talents emerged, and he soon became a successful businessman. He started a scientific supply house, Southwestern Scientific Supply. In large measure, his immediate success related to the huge problem in the availability of leopard frogs from the Midwest and New England. Companies in Wisconsin and Vermont that normally sold frogs to schools, universities, and laboratories could not meet the demand, because there were large drops in leopard frog populations during the late 1960s and 1970s. More than likely, this was due to chemical poisoning attributable to fertilizers and pesticides that polluted the waters. Lee took advantage of a discovery he had made in Mexico during his graduate student days. Near Los Mochis, Sinaloa, there is intense agriculture activity based upon an extensive irrigation system fed by many canals. This great availability of water created a superb habitat for local leopard frogs, leading to immense populations that could be easily harvested. The ready availability of frogs and cheap labor put Lee in business and insured success. I was pleased that he was doing so well, but I was bothered by the fact that he was selling his leopard frogs as *R. pipiens*, the common American species supplied by the classic supply houses. I knew that the frogs from Sinaloa were not *R. pipiens*, but another species, perhaps one that had not yet been described. I called Lee on it, and we wrote a short paper on the subject essentially pointing out differences between his "Mexican *R. pipiens*" and the American species. At this time, a young man who had a degree in biology was working for Lee. When he learned about the discrepancy between the frogs that Lee sold and those from Wisconsin or Vermont, he became excited enough to want to work on the problem. Thus, this bright young man, John Frost, applied to our graduate program, and in due time he became one of my more accomplished graduate students.

I had no great practical reason to be concerned with this leopard frog challenge, but early on I did depend upon Lee's company to provide me with a small number of Mexican leaf frogs that were also abundant in the Los Mochis area and that could be caught easily by the *raneros*. My use of these adult frogs was to try to get them to breed and to utilize them as a basis for a breeding colony while we were working on the dermal chromatophore unit. We found that these frogs had a bizarre, large and compound type of melanosome, the cortex of which was composed of the unknown pigment that we were studying with Peppe Prota. Since we knew that the center of this melanosome was composed of eumelanin, the usual melanosome pigment of most vertebrates, I was very much interested in learning how the leaf frog melanosome developed. To this end, I would need embryonic and early larval stages of the frog. I could not depend upon Lee and his *raneros* to collect these early stages and to get them to Tucson with the speed that was needed. Since we could not go to Sinaloa, I thought that perhaps I could bring Sinaloa to us, in particular a hot and wet Sinaloa typical of the summer rainy period.

It occurred to me that this could possibly be accomplished by purchasing a small greenhouse kit that could be assembled outdoors in a safe place that had access to water and electricity. The university farm on Campbell Avenue turned out to be just the place. I had another idea about supplying natural food in a simple way. If a UV light were installed behind a chicken wire-covered window, an abundance of moths would be attracted to and into the greenhouse providing a super abundance of natural food for the frogs. To provide cover and to help maintain high moisture levels, we installed a number of plants in large pots and a plastic tub to provide a water source. The last problem to be dealt with was the intense heat that was bound to accumulate in this plastic enclosed structure sitting out in the Arizona sun. Somewhere we found a small evaporative cooler that both solved the high temperature problem and helped maintain the humidity.

The university greenhouse worked fine, and we kept it in operation for several years. It was so successful that I erected another at my own home that worked even better, because we had a better supply of insects. Since our home is in the lush desert near a large wash, there were many more moths available, and spontaneous breeding was more reliable. Over the years, the two greenhouses provided a continuous supply of eggs, tadpoles, and young adults that were used for a variety of studies. Every now and then, a few new adults were added to the colonies as John Frost brought them from other parts of Sinoloa and Sonora. John collected them as he did field studies in these parts of Mexico as part of his graduate work on the leopard frogs of these areas.

The common origin of pigment cells

Notwithstanding the attention that I was now giving to the Mexican leaf frog, my activities were also moving in other directions. One of these was the concept that all the basic types of chromatophores, or pigment cells, are very much related with respect to cellular and organellar origin. Of course, it had long since been known that vertebrate pigment cells are derived from the neural crest, and that these prospective pigment cells migrate from this site of origin to populate all areas of the surface of vertebrate organisms and internal locations as well. From the observations made by John Taylor at the EM level in his work in our lab and subsequently, it appeared that each of the pigmentary organelles, including melanosomes, reflecting platelets, and pterinosomes, all start out as a small vesicle pinched off from the intracellular membrane system, the endoplasmic reticulum. With this knowledge in mind, we began to better understand the significance of a series of studies we had made during the late 1960s and 1970s. One of these was related to a curiosity of mine about the various color varieties of a charming little salamander, the red-backed salamander, *Plethodon cinereus*, that was common in areas in and near Rochester, where I grew up. On one of my trips to Rochester in the late 1960s, I collected a few individuals on the banks of the Genesee River, and John Taylor and I did a little collaborative study. He was then an assistant professor at Wayne State University. We found that the red pigment of *P. cinereus* was the pteridine pigment, drosopterin, and that this pigment was found in organelles that also contained melanin. Thus, two different, unrelated pigments, each associated with its own organelle, were here found in one compound organelle. This was a startling discovery.

At about the same time, Jiro Matsumoto, who was then in my lab, made a related observation on the red dorsal stripe of a garter snake (*Thamnophis*). Here he found a layer of pigment cells that contained both pterinosomes and reflecting platelets within the same cell. These were truly mosaic pigment cells. Perhaps the most significant work in this series of studies came from an examination of the iris of the local Inca dove, *Columbina inca*. Our colleague, Robert Chiasson, was engaged in a study of why the doves developed bright red eyes when they were excited, and there was the possibility that some sort of reflecting cell in the iris played a role. I obtained some of these ground doves, and together Sam Ferris and I discovered that, indeed, the iris contained pigment cells very much like the classical iridophores of lower vertebrates. These avian iridophores actually were mosaic cells in that they contained both melanosomes and reflecting platelets. Of even greater significance was the fact that some of the organelles were mosaic such that both melanin and purine pigments were present within the same organelle.

Altogether, these observations of mosaic pigment cells and, indeed, mosaic organelles provided seed for the thought that all pigment cells arose from a kind of stem cell of neural crest origin. As time went on, we made other observations on the pigment cells of the Mexican leaf frog that supported this view. In 1977, I put together all these observations with those available from other studies on mammalian species, and developed an hypothesis that all pigment cells are derived in a similar way from a stem cell of neural crest origin. I drafted a manuscript to describe my idea, and it was accepted and published as a lead article in *Science* magazine. This paper probably represented one of my more important research contributions. Unfortunately, it falls into the realm of unfinished business, for while the hypothesis gained wide acceptance and approval, no one has followed through with any further studies that would offer definitive proof to support it.

Photosensitive chromatophores and the Mexican leaf frog

The arrival of the first shipments of leaf frogs that Lee sent me from Southwest Scientific Supply presented a new and interesting color change phenomenon. In anticipation of their arrival, we made a simple cage of chicken wire that I placed outdoors and adjacent to my amphibian facility on the roof of our biosciences building. The frogs were delivered in a fairly large cardboard box that I opened and tossed into the cage to allow the frogs to emerge at their convenience. When they did, they immediately assumed their daytime pose in which they flatten themselves against a substrate, close their eyes, and fold their legs tightly against their body. The dorsal skin is impervious to water, and thus this pose prevents the frogs from drying out. When the frogs arrived, they were a rather light brown in color, probably some sort of color adaptation to the shipping container. At any rate, the surface of some of the frogs was partially covered by the box after they emerged and assumed their daytime pose. Later in the afternoon, when I lifted up the box to dispose of it, I was very much surprised to find a sharp line of color demarcation between the covered and uncovered parts of the dorsal surface. That part that had remained uncovered was still a brownish color whereas the covered part was now green. The sharp line that separated the two shades of color coincided exactly with where the edge of the box had been. This observation immediately suggested that the skin or some part of it was directly sensitive to light. I was impressed enough that, during the next few days, I did a few primitive experiments to refine this preliminary observation. I simply cut out some masks from black plastic and placed them on the dorsal surface of a frog. Some time later, I removed the mask to see what happened. When I used cut outs of U and A placed side by side to mask the surface, I obtained a U A frog with the letters green and the surrounding area tan. No doubt the skin possesses a photosensitive capacity. I was too busy to follow this lead immediately; it would have to wait its turn.

The wait was not long. In the autumn of 1973, Tetsuro Iga, a young investigator from Shimane University in Matsue arrived in my laboratory to work on chromatophore physiology. He set to work on the leaf frog skin photosensitivity phenomenon that I had discovered. His initial experiments revealed that the chromatophores were not innervated, so the effect was unlike that which had been disclosed for chameleons and other lizards. Additional studies gave no further insight into the basis for the phenomenon, and we could see no easy way to solve the problem. Accordingly, the explanation of the integumental photosensitivity of the leaf frog became another example of unfinished business as far as my laboratory was concerned. In a broader sense, this is another example of the unexplained phenomenon known as the "dermal sense" that occurs in a variety of lower vertebrates and that no one has elucidated.

Some evolutionary considerations about the phyllomedusine melanophore

After our return from Naples where preliminary studies on the chemistry of the red pigment from the Mexican leaf frog were underway, I waited with great anticipation for word about its identity. In the meantime, I pursued some biological aspects of the red melanosomes. First, we needed to name its pigment, and we came up with "rhodomelanochrome" since it is a red pigment contained in a melanophore. We first used this name in an article published in *Science* in 1973. Soon afterwards, Sam Ferris and I described the exact location of the pigment within the melanosome. As part of this work, we devised a simple test for the presence of the pigment based upon its solubility in NaOH.

Right from the start, I was very much taken with the resemblance between the Mexican leaf frog and White's tree frog (*Litoria caerulea*) from Australia. None of the American species of tree frog (hylids) that we tested had rhodomelanochrome. I wondered if perhaps White's tree frog might be more closely related to the New World leaf frogs (*Phyllomedusinae*), so I set about getting skin specimens from it and from related species found in Australia and Papua, New Guinea. I was able to obtain skin from only five species of the genus, and unfortunately they had been fixed in formalin and stored in alcohol. Nevertheless, it was possible to test them with NaOH, and I found that two of the species tested positively for rhodomelanochrome and three were negative. Despite the poor fixation, skin from two of the positive testing species and from two of the negative ones were examined by electron microscopy, and Sam Ferris was able to provide useful images. The two species positive for rhodomelanochrome, *L. caerulea* and *L. infrafrenata*, both possessed the large fibrous compound melanosome typical of the six new world phyllomedusine species that John Taylor and I had examined. The two negative species, *L. nasuta* and *L. peroni*, possessed the smaller, simple eumelanin type of melanosome typical of most other vertebrates. These were remarkable findings, because they suggest that at least some of the green tree frogs of Australia have an origin in common with the phyllomedusine species of America. I wrote about our results in a manuscript for the respected herpetological journal *Copeia*. In it, I suggested that this common origin occurred in the late Triassic before the separation of the southern landmass of Gondwanaland. During subsequent continental drift, loss of the Antarctic connection between South America and Australia could have led to the separation of the closely related species of the phyllomedusinae and those of the genus *Litoria*. It turned out that *Copeia* found this speculation of interest, for the paper was published in 1975. For my part, this business was finished, and the ball was now in the court of the herpetological world. I have not followed the literature in this area, but I have learned from herpetologist friends that current revisions of amphibian taxonomy have taken into account my results of more than thirty years ago.

Leopard frogs and Arizona

Concurrent with the attention given to the Mexican leaf frog in the early 1970s, much was going on in my lab concerning leopard frogs. John Frost began working hard from the first day he joined the lab. He was an independent thinker and a capable worker, and while I provided intellectual, moral, and financial support, John went ahead on his own. An important contribution that I made was to show him how to deal with the reproductive biology of frogs and how to grow and maintain a frog colony in captivity. He put some of this new knowledge to work as he moved from working for Lee Stackhouse to undertaking the broader subject of leopard frogs in Arizona and northwest Mexico. I had accrued a lot of information about leopard frogs in Arizona just because of my own personal interest. This helped provide a good start for John.

The first question was, "What are the frogs that *raneros* were harvesting in Los Mochis?" The predominant species that made up about ninety percent of the frogs collected was a large leopard frog that the literature revealed to be an already described species, *R. berlandieri forreri*. The other was a new species that we described in 1976 as *R. magnaocularis*. As the name implies, a characteristic feature of the frog is disproportionately large eyes. Otherwise, the frog resembled one of the undescribed species of Arizona leopard frogs that was common in southern Arizona. It was then referred to as the lowland form. As part of his PhD dissertation, John did an analysis of why *R. berlandieri forreri* and *R. magnaocularis* were reproductively isolated.

When I arrived in Tucson in 1956, there was not that much interest in leopard frogs, but herpetologists at least recognized that in southern Arizona there existed two distinctly different types, probably different races of the common American leopard frog, *R. pipiens*. One of these was large and had rough skin that was sometimes dark enough to almost obliterate the leopard spots. It was especially observed in the Chiricahua Mountains by naturalists frequenting the Southwestern Research Station that is located there. This frog came to be known as the southern form. The other leopard frog was found at lower elevations and was widely distributed in many of the canyons and streams not far from Tucson. It came to be known as the lowland form. Before John arrived, I had learned much about when and where these two forms bred and was able to use their embryos for my own research or for teaching purposes in my two embryology courses. The embryos of the two species were certainly different, and this was especially true with respect to temperature tolerance. Embryos of the lowland form were quite tolerant of cold temperatures, not unlike those of the northern leopard frog, *R. pipiens*. In sharp contrast, embryos of the southern form were quite sensitive to low temperatures. Perhaps this explains why the southern form normally bred late in the spring or early summer when the ambient waters were warm.

I showed John where I had found both types of frog, and he began to explore and found many more locations. Since I enjoyed being in the field, I accompanied him on many, many frog excursions. On one of these trips into Turkey Canyon on the west side of the Chiricahua Mountains, we found an unusual tadpole marked by a red stripe on its belly. We had no idea what it was, so we took some tadpoles back to the lab, and John raised them through metamorphosis. To his surprise, it turned out to be *R. blairi*, a species common in Texas, but never before found in Arizona. John followed up this discovery and learned that the species was limited to very southeast Arizona where it thrived in the irrigation ditches of this agricultural area. Obviously, this Arizona population was a relic of earlier times when *R. blairi* had a much wider distribution in the southwest.

One of our field trips was interesting and included an amusing experience. I knew that the lowland form thrived in water sources fed by hot springs, and I had long been curious about Hooker Hot Springs. This site was on a remote ranch that had been a famous resort in the early 1900s, but had reverted back to a working ranch. It was located in a lush desert east of the San Pedro River, a bit north of Benson. We decided one morning to head out toward Hooker Hot Springs, first checking on a series of desert stock tanks *en route*. By the time we reached the ranch, we were pretty grimy. There was no one in the ranch yard, so we knocked on the ranch house door. After some delay, the door opened, and there stood a very attractive young woman in her mid-twenties, the rancher's wife. We detected her great apprehension, but fortunately John Frost made a wonderful first appearance, and she granted us permission to investigate the hot springs. She must have been at home alone and to be confronted by two raunchy and virile looking men in this remote place must have been a shock! In more recent times, The Nature Conservancy purchased the ranch and refurbished it. It is now known as Muleshoe.

John's work was being recognized at about the same time that herpetologists were becoming interested in Arizona frogs. One of these scientists, John Mecham from Texas Tech University, felt that possibly the southern form was a distinct species, as did John Frost, so he asked John to collaborate with him to describe it as a new species. They did so under the species name *R. chiricahuensis*, a name that I thoroughly disapproved of for several reasons. I think that John Frost agreed with me, but he had no say in the matter. At the same time, a young investigator from Arizona State University, John Platz, asked John Frost to collaborate with him in describing the lowland form as a new species. He selected the name *R. yavapaiensis* because he was working with populations from Yavapai County. I found this name even more deplorable for even more compelling reasons!

Late in the summer of 1973, with great fear and trepidation about the financial risk, Lou and I bought a cabin on three-plus acres in the village of Greer in the White Mountains of eastern Arizona. With both of us working in those days, we could not spend much time enjoying the cabin and the exposure to nature that it afforded. Nevertheless, I found enough time to observe the two leopard frog species that abounded there in those days. The northern leopard frog, *R. pipiens*, was present in numbers right in Greer, and *R. chiricahuensis* was present only a few miles away. When John joined my lab, we went to the cabin together on several occasions to collect and observe frogs. Eventually, he used these White Mountain frogs in crosses between and among the various other frogs of southern Arizona and Mexico. The data that derived were important in his post-dissertation research.

Soon after his initial visits to our cabin, John visited it on what must have been certainly a pleasant occasion. A year or so after he joined the lab, I acquired a new graduate student, Sally Viparina. She was interested in modern embryology, just then being referred to as developmental biology. Actually, she was to have done her work at Purdue University with one of the new breed of developmental biologists; however, she was in a serious automobile accident that injured her badly and required a long recovery. In addition, her faculty advisor died unexpectedly. She wrote to me to ask if I would accept her as a PhD student, and so she came to Tucson where she continued to heal in our warm climate. It didn't take too long before she and John Frost fell in love and in due time decided to marry. Lou and I regarded both of them with affection, and so we were pleased to serve as best man and matron of honor when they were married. We offered them our cabin in Greer for their honeymoon. They were soon back in the lab, each working in entirely different areas. John continued his work on frogs, and Sally took up working with pteridines, primarily in the Mexican leaf frog.

Identification of rhodomelanochrome as pterorhodin

About the time that Sally began working on *Pachymedusa* pteridines, a breakthrough occurred with respect to the identity of rhodomelanochrome. Peppe wrote that at last he thought that the red pigment was pteridine in nature and that it would help if he could obtain some pure, authentic pteridine samples. The authority on pteridine chemistry at that time was Professor Max Viscontini of the University of Zurich. I wrote to him to ask for samples and to explain why we needed them. He replied that he had read our 1973 paper in *Science* and he knew from the spectra we showed that our rhodomelanochrome was actually a pteridine dimer, pterorhodin that he had worked with. He sent authentic samples of pterorhodin that would allow us to verify the identity for ourselves. He said that he did not contact us at the time we published the paper in *Science* three years earlier because he knew we would eventually figure it out for ourselves. My immediate reaction to his letter was, "Oh, how typically Swiss," but at the same time I was grateful to have finally learned the identity of the mysterious red pigment of the phyllomedusine skin. This knowledge was significant from several standpoints. First of all, pteridine dimers are not common in nature. They had been known only from nereid worms and from the eyes of some species of drosophila. The presence of this dimer pterorhodin in the skin of leaf frogs marked the first time that it had been found in a vertebrate species.

Even while we were waiting for the chemical elucidation of our red pigment, we made good progress with respect to important aspects of its biology. One of the more elegant pieces of work coming from my laboratory was the result of a collaboration I had with Sam Ferris, and John Taylor and his student at Wayne State, Bill Turner. We generated a paper published in *Developmental Biology* in 1978 dealing with melanophore differentiation in leaf frogs. When we first observed the large compound melanosome of the Mexican leaf frog, we found that the dark internal kernel that it possessed was eumelanin, so typical of vertebrate melanosomes. Later, as we began to look at the skins of tadpoles, we found that their melanosomes were like the typical vertebrate melanosome except that they were much smaller and, in fact, were about the same diameter as the kernel of the adult red melanosome. This fact suggested that perhaps the tadpole melanosome was transformed into the adult type during metamorphosis. Accordingly, we examined the melanosomes at stages from late tadpoles to transformed froglets. At the same time, we tested each of these stages with NaOH. We found that the melanosome of tadpole stages was enclosed by a limiting membrane tightly opposed to the melanin surface. These stages tested negatively for NaOH. At a specific stage, well into metamorphosis, small flocculations of material began to appear in spaces that were forming as the limiting membrane began to uplift. These stages just barely tested positive with the NaOH test. In successive stages, more of the material was laid down in the space and, by the end of metamorphosis, the kernel was surrounded completely by a concentric mass of fibers. The NaOH test revealed more of the red substance with each successive stage. With the discovery that the red pigment was pterorhodin, the question arose, "What is the mechanism called into play that brings about the synthesis and deposition of the pteridine dimer at these specific stages of development?"

With the knowledge that every specific event in amphibian metamorphosis is controlled by thyroxine, it seemed reasonable to attack at this level. One approach was to do what I had done with *Pleurodeles* years earlier, namely to implant thyroxine/cholesterol pellets locally under the skin of pre-metamorphic stages of *Pachymedusa*. For logistic reasons, we were not able to obtain local metamorphosis. Another approach was to graft early metamorphic stages of larval skin onto froglets. The hope was that the graft would respond to the post-metamorphic environment by prematurely producing pterorhodin. This did not work either. Sally Frost was working on pteridine metabolism in *Pachymedusa*, but this did not provide any insights other than her discovery that pterorhodin production could be blocked by allopurinol. The latter blocks the enzyme xanthine dehydrogenase, a key enzyme in pteridine synthesis. About this time, another Japanese postdoctoral associate, Hiroyuki Ide, came to the lab to confront the problem through cell culture. In Japan, Hiro had successfully cultured bullfrog melanophores and now tried his hand with *Pachymedusa*. He was able to culture larval melanophores, but never got them to transform and produce pterorhodin or to develop the post-metamorphic form. A few years later, one of Hama's students from Nagoya, Masumi Yasutomi, came to try his hand with the cell culture approach. He did a nice little study, but he could not get the adult type of melanosome to transform in culture. Thus, again I was faced with a piece of unfinished business that probably will never be finished. It is an esoteric problem that may not have appeal in these modern times, but is nevertheless important.

In gratitude to John and Sally

Much of the work going on in my laboratory during the mid to late 1970s was based upon the successful establishment of a breeding colony of Mexican leaf frogs. It produced many of the embryos and larvae employed in a variety of studies. While I did most of the overseeing of the colony that was housed primarily at my home and on the university farm, I very much depended upon John and Sally in different ways. John helped with the construction and maintenance of the rearing facilities and, in fact, built a large structure at the university farm where he raised his various leopard frogs and where I also kept part of my leaf frog colony. This structure, "Rancho Rana," was an elegant and effective affair that demonstrated very well John's talents as a carpenter. During those years between 1974 and 1978, I depended heavily on John to take care of things. At the same time, Sally was more or less in charge of the lab. She kept things organized and going. During our exposure to French culture, Lou and I learned of the reputation of Bretonne housewives with respect to maintaining the home and family. In effect, Sally was the tadpole patrol's Bretonne wife. With the two of them on hand, we felt comfortable to be away. They would house sit for us and at the same time take responsibility for the amphibians and for the lab.

John finished his PhD in 1976 and stayed on as a post-doctoral student while Sally finished her dissertation work in 1978. In the fall of that year, Sally took a post-doctoral fellowship at Indiana University, and John went along as a kind of adjunct. He soon took advantage of his reputation as a scholar of leopard frog evolution and obtained support from the National Science Foundation to continue the work that he had been doing in my lab. Sally and John left Indiana when Sally was offered an assistant professorship at the University of Kansas. Again, John went as an adjunct with continuing support from NSF. They each had great success. Unfortunately, they experienced marital difficulties that led to divorce, and John left Kansas to start a new life in North Carolina.

Sally's scholarly efforts at Kansas gradually took on a new dimension as she became involved in university administration. In time, she became Dean of the College of Liberal Arts and was so successful that she moved on to become Provost at Purdue University. During these developing years, she remarried a Californian, Ken Mason, with whom she collaborated in work on the molecular aspects of pigmentation. When they moved to Purdue, Ken joined the faculty and pursued his own career. In August of 2007, Sally, now Sally Mason, was named President of the University of Iowa. Needless to say, Lou and I were delighted and proud of her success. We followed her recruitment closely through our friends at the University of Iowa Foundation, the fund-raising arm of the university. They were kind enough to give us frequent updates on the process. It is remarkable that Sally, my PhD student at Arizona, became President of our *alma mater*, the University of Iowa. Our initial pride has been enhanced during these subsequent four years during which Sally has become an exceedingly capable president during especially trying times. We have been pleased to be her guest in Iowa City several times during this period.

12

Important Travel in 1977

To have had John and Sally available to shoulder some of my responsibilities during 1977 made it easy to carry out some travel obligations during this year. The first trip was not a long one, but it became significant to later developments in my life. In the spring of the year, the 8th International Congress of Developmental Biology was held in Tokyo. I was a speaker at two sessions. The first concerned the pleuropotentiality of the neural crest and the second described the developmental biology of the phyllomedusine melanosome. While these presentations and meeting with colleagues to discuss problems in developmental biology were the purpose of my presence in Tokyo, the true significance of the stay in Tokyo turned out to be something different, thanks to my friend Jiro Matsumoto. One day early in the Congress, Jiro appeared and aside from just saying hello, he brought with him a special invitation to tea at a cancer institute. The invitation was from Prince Masahito, the second of Emperor Hirohito's sons. The events leading to this invitation are an interesting story.

Jiro's first sabbatical started in my laboratory and his second was with John Taylor at Wayne State University in Detroit. He, Akiko, and their daughter, Mina, had a great year in Detroit helped partly because of the year that John and Linda had spent with Jiro and Mat in Japan. As John's program developed at Wayne State, he had gotten involved in the cell culture of the red pigment cells (erythrophores) of goldfish. Jiro became involved with this work during his stay at Wayne State and, as usual, with great success. When he returned to Yokohama, he continued the work. Jiro tells that one evening when he returned home from work, the telephone rang and to his astonishment the caller was Prince Masahito. All of the royal family had interests in biological sciences, and Prince Masahito was a patron of the important zoo and aquarium in the large Ueno Park. One day when the Prince was visiting the aquarium, he was informed that some of their goldfish were afflicted with red cancerous growths on their skin. He was curious and decided to follow up through another of the royal family's connections, one of Tokyo's cancer institutes. Biopsies of these red lesions revealed that indeed they were cancerous and that in fact they were erythrophoromas, the red pigment cell equivalent of melanomas. This led to the contact with Jiro Matsumoto who was invited to come to work on the erythrophoromas at the cancer institute. Thus, Jiro began to culture these red cells in collaboration with Prince Masahito and others at the institute. In the process, Masahito began to learn about pigment cell biology. By this time in my career, I was well known as a pioneer in the field, and when Masahito learned from Jiro that I was to be in Tokyo, he indicated that he would like to meet me. Thus, the invitation was extended through Jiro for me to come to the cancer institute for tea. Needless to say, I was thrilled and honored to meet Masahito and his colleagues. We had a very pleasant session. This was my first opportunity to meet with Prince Masahito, but not my last.

In the meantime, the tadpole patrol activities continued in Tucson where Sally discovered a rather serious problem. She called Lou about it, and Lou telephoned me in Japan to relay the story to me. It seems that Sally had gone to University Stores to pick up some supplies. I had an account there funded by one of my research grants. In the process, she found that someone other than a member of the tadpole patrol was picking up supplies and charging them to our account. She launched her own investigation and found that the culprit was the head of our department, who was not authorized to use our funds. By the time I returned to Tucson, Sally had told Mac Hadley what had taken place, and he in turn found that his account at stores had been a similar victim of the head. My plan was to stop this activity by going to A. R. Kassander, Vice President for Research; however, Mac became so enraged that he either threatened to report or actually did report the problem to the United States Public Health Service, from which his research funds were derived. In any case, his action led to quite a brouhaha that eventually calmed down. If it were not for Sally Frost's vigilance, none of us might have known about our victimization.

The next trip was for a prolonged stay in Germany. One of my friends from the international world of endocrinology was Professor Wilfried Hanke who had just become head of one of the zoology departments at the University of Karlsruhe in the Rhine Valley of Germany. Wilfried asked if I would be willing to teach a special six-week course on pigment cell biology in his department (Lehrstuhl II). I would have a special appointment as Visiting Professor. Of course, I jumped at the opportunity, not only because I was free from my own teaching, being on a short sabbatical leave at the time, but because I would have the chance to polish my use of the German language, even though the course was to be given in English. Nevertheless, I anticipated speaking German to a considerable degree, every day. The course was to start in mid-November and would terminate just before Christmas. So off I went, leaving Lou behind at her work as a travel agent. The plan was for her to come to Karlsruhe for two weeks as my course ended.

I arrived at a beautiful time in the Rhine Valley for the temperatures were still mild, and the trees still bore their colors of autumn. After a warm welcome, Wilfried introduced me to his graduate students, including Barbara Stehle who was to be my teaching assistant. This was a surprise, for I already had met Barbara, who went by her diminutive name Bärbel, in Berkeley where she had spent a pre-doctoral year with our friend Howard Bern. Lou and I had gone to Berkeley to visit Howard and Estelle and to give a seminar. Howard had a party at his home one evening, and Lou and I spoke with Bärbel at some length. Little did I know that we would meet again in Karlsruhe. While it was an advantage to have my assistant speak perfect American English, it was a real disadvantage toward my hopes to learn to become a fluent German speaker. To compound the situation, all the graduate students were fairly proficient in English, and they wished to improve by speaking only English with me. A few years earlier, when I worked at the *Stazione* in Naples, I had been much impressed by the librarian there, Walter Gruben, who was fluent in English, French, Italian, Spanish, and German, among other languages. One day I asked Walter how it was that he was able to speak so many languages. He explained that when he escaped from home in the Baltics during WWII and moved from one country to another, he always sought out a "long-haired dictionary," meaning a local girlfriend, and thus, by total immersion in a language, he learned. Unfortunately, despite my almost total immersion with the students, there was little immersion in their language.

It is amazing how close I became to the students and how kind they were to me. I was placed in a special position between them and their mentor, Professor Hanke. It was a bit awkward for me, but I understood the situation very well. Although I was the professional equal of their professor, I could become their friend, unlike Wilfried Hanke who was bound by German academic tradition that kept him several arm lengths away. He was not a severe taskmaster; rather, he was a mild and pleasant person. Perhaps because of my friendship with Bärbel, I was frequently asked by the students to join them in some of their social gatherings. One of the students, Nana, was a particular friend of Bärbel's, and she and her boyfriend, Andreas, organized some of the events. They lived together in a housing unit that was kind of a super dormitory. It had a common kitchen where the students could prepare their own meals.

One gathering was particularly interesting and pleasant. Nana and Andreas invited me for dinner and asked me to go with them to procure the main course, *forelle* (trout). We drove off to the nearby Reingau, an area just north of Karlsruhe, defined by a bend in the Rhine. It was an agricultural area noted for its vineyards. We arrived at a facility that raised trout where one could fish for himself or purchase fresh trout that the proprietors had just caught. Andreas chose the latter, and we left with a substantial package of fish. A little farther down the road, we stopped at a winery where Andreas knew the *winzer* (wine maker). We purchased a few bottles of their white wine and headed home. Of course, we were in need of sustenance by then, so as we passed through a pleasant forest, we stopped at a small facility that was essentially a snack bar. It featured *hausmache*, an array of home-prepared items such as hams, sausages, salamis, and various other kinds of preserved meat products and cheeses. There was coarse country bread to go with it all. Lou tells me that I remember things like this because I am a foodie, and perhaps I am. When we got back to town, I was dropped off while Nana and Andreas returned to the dormitory to prepare the evening meal. Later, Bärbel picked me up, and we went to the dormitory to join Nana and Andreas, another graduate student couple, and a special guest, Wilfried

Hanke. We were having a fine time in anticipation of the meal that Nana and Andreas had prepared. The featured dish was the *forelle* that Andreas had baked in his own special way. Each of us was served a trout wrapped in foil. Of course, all was delicious, and we had a wonderful evening. Wilfried, I know, enjoyed himself in this wonderfully relaxed ambiance. While I am expansive about the hospitality shown me by the graduate students, I do not mean to ignore that shown me by Wilfried Hanke and his wife. I was invited to dinner at their home on several occasions. Later, when Lou came, we enjoyed a lovely evening at their home. Often, on Sundays, Wilfried invited me to take long walks with him, rain or shine, in the typical German fashion.

My friendship with Bärbel became special for several reasons, in part because of the California connection and because we spent a lot of time together in teaching my course. She was also lonely, because her fiancé was away working in northern Germany. Most important of all was that I came to know her family very well. They lived in a small town not far away. This was Advent season, and on Sunday mornings, Bärbel took me to have a festive breakfast with her parents and two brothers. Her parents were very nice, and I enjoyed speaking with her father. His English was quite good. He was principal of a gymnasium in nearby Rüdesheim, an important town on the Rhine. Unfortunately, I met Herr Stehle only once, for a few days later Bärbel informed me that her father needed to go to Mainz for some medical tests. The results revealed that Herr Stehle had a malignant brain tumor that required almost immediate surgery. This was done within the week, and sadly, within a few days, he lost this final battle. What a catastrophe it was for him, a vital and vibrant person, to perish within a week or two. I suppose that this sad event was another factor in cementing the friendship that I developed with the Stehle family.

My course went on without Bärbel for a few days. By now, I was firmly entrenched in the department, and aside from the course I taught, the graduate students would come to me to discuss their research. I was not so much a source of help; instead I was a knowledgeable person who would listen as they told about their work. Bärbel discussed her research project with me, and in this case I was able to offer help. Her research project concerned the effects of prolactin on *Xenopus* tadpoles, and she was about to inject this hormone into these fragile larvae. She was faced with the problems of mortality, hormone leakage, and the stress of frequent injections. I well appreciated the problems she faced because of my long experience of injecting hormones into the much hardier tadpoles of leopard frogs. I suggested a better way of getting prolactin into the *Xenopus* larvae. It was actually a technique that I had taught my graduate student, Norma Briggs, for use in her dissertation which she had completed earlier in the year. It was based upon the fact that prolactin is one of the hormones of the anterior lobe of the pituitary whose release into the circulation is regulated by inhibition rather than by stimulation. In other words, an anterior lobe separated from the brain would spontaneously release prolactin. The procedure was that an isolated *pars distalis* be implanted into a *Xenopus* tadpole in a place where it would not be disruptive and where it would be supported and nurtured by the host. Thus it would continuously release prolactin, insuring a constant release of the hormone into the host's circulation. The space between the external carotid arteries in the ample lower jaw of the larva was an excellent and effective location. The disruption in her life associated with her father's death, together with the onset of the Christmas recess, precluded Bärbel's doing much of this work before I returned to Tucson. Early in the new year, she sent me photos of the implanted region and kept me posted on her success.

The Christmas season was rapidly approaching. About two weeks before the end of my stint in Karlsruhe, Lou joined me for a bit of a vacation and for a taste of Germany during the holiday season. We had no means of transportation of our own, but between Bärbel and my student friends Nana and Andreas, we did very well. The latter took us on a lovely outing on some beautiful roads in the *Schwartzwald* (Black Forest). Bärbel's fiancé, Egbert Kahl (Eggie), had returned from the north, and the four of us spent a festive Sunday and overnight in the nearby important city of Mainz where Eggie's parents lived. His father, Gunther, was an art historian and was a scholar of the Mainz cathedral. He kindly gave us a tour which we enjoyed almost as much as the *Weinachts markt* (Christmas market) that filled the large square near the cathedral. The *Weinachts markt* dwarfed the one in Karlsruhe that we had come to know very well. The latter was just across the main road from our building at the university, and practically every day while Lou was there, we walked over to it for lunch. The season was still mild, and it was great fun to partake of a *brötchen* and *Thüringer wurst* (hard roll and sausage) at one of the stalls. Of course, this had to be washed down with a glass of *gluwein* (a hot, spiced red wine) typical of the season.

A few days before Christmas, we returned home joyfully. It had been an eventful and fascinating experience, but six weeks away from home and my Tucson life were beginning to wear on me. I bade farewell to Bärbel and her family, but a lifetime friendship had been forged.

13

The Emergence of Pigment Cell Biology

Thus far, I have referred to pigment cell biology without really defining the term or discussing its origin. Its definition is not easy, but in its simplest form it is essentially all there is to know about pigment cells, their form, functions, structure, composition, origins, physiology, pathology, and their variations, as they exist in living forms from invertebrates, to lower vertebrates, to higher vertebrates, including humans. These cells are all related by either homology or analogy. The concept developed as investigators sought answers to specific questions that arose during their studies. It became solidified as they sought help from colleagues in other disciplines.

Pigment cells have long fascinated biologists and lay people alike, because their expression provides the striking variety and array of colors and color patterns of animals from primitive invertebrates to man. Thus, they become subjects of study for investigators from a variety of disciplines. Early on, chemists and physicists began their attempts to elucidate the chemical and physical nature of animal pigments, and they made much progress as techniques and procedures evolved during the late 1800s and early 1900s. By the beginning of WWII, an amazing sophistication about the physical and chemical foundations of animal coloration was achieved. Two landmarks of importance were: 1) an understanding of structural colors, in particular, Rayleigh or Tyndall scattering; and 2) the discovery of mushroom tyrosinase, a key enzyme in melanin synthesis.

Concurrent with the growth of knowledge about physical and chemical aspects of pigmentation, an understanding of the cellular basis of pigmentation began to emerge. As was mentioned earlier, it was known rather early that some pigments are found in cells; in fact, the word chromatophore seems to have been derived from the Italian word *chromoforo* employed by Sangiovonni in 1819 to describe the pigment-containing sacs of cephalopod chromatophore organs. An understanding of the nature of true chromatophores was derived mostly from a series of works of the early 1900s carried out by German workers on fishes and amphibians. Cytological details about these cells came slowly in the absence of the electron microscope; however, an overwhelming array of physiological work was done, especially with respect to color change phenomena. As the 1900s proceeded, investigators in other disciplines concerned themselves with pigment cells. Developmental biology (embryology) was a dominant field, and so, some investigators devoted special attention to the embryonic origin of pigment cells. Thus, in the mid-thirties it was revealed that amphibian pigment cells were derived from the neural crest. Later, it was shown that pigment cells of birds have a similar origin, and subsequently, the avian system was used to demonstrate that pigment cells of mammals are also of neural crest origin. This period was marked by an interest in the development of pigmentation patterns. Such activity was stimulated by and went hand-in-hand with progress in genetic research. The decade of the 1960s was an exceedingly important one for the genetics of pigmentation. It was during this time that a large number of mouse pigmentary mutants were discovered. As time has gone on, these mutants and many new ones have contributed mightily to the growth of pigment cell biology in these modern times.

The fact of the matter is that it is a genetic phenomenon that provided the impetus for the formalization of the discipline of pigment cell biology. This is how it transpired. In 1946, inspired by the urgent needs of physicians to comprehend the nature of human melanomas and his own need to understand the biology of melanoma that arose in hybrids between swordtail fish and platyfish, Professor Myron Gordon championed what became the first International Pigment Cell Conference (IPCC). In his own words, "It was decided to ask workers in the biological, medical, physical and borderline sciences to meet together, to analyze the nature and behavior of pigment cells and their relation to melanoma development. The formation of an integrated group of specialists was visualized who would summarize the available facts and who would thereby establish a pool of information readily accessible to all workers." Thus the tone and objectives were set for a series of International Pigment Cell Conferences that were held every three or so years and have continued since 1946. The first few of these conferences were organized by Professor Gordon and were sponsored by the New York Zoological Society. Myron Gordon passed away in 1959, but thanks to the initiative, energy, and industry of Dr. Vernon Riley, the IPCCs persisted, not in New York City, but in other venues.

Although the first five conferences were considered international, they were all held in the United States. The first IPCC to be held elsewhere was organized and held by the International Union Against Cancer in Sofia, Bulgaria, in 1965, with Dr. Riley serving on the conference program committee. Despite a heavy emphasis on melanoma, this conference retained its continuing philosophy of bringing together researchers from diverse disciplines. Again, through the organizational capacities of Dr. Riley, the conferences returned to the United States, and the VIIth IPCC was held in Seattle in 1969. At this conference, first indications of a future permanent organization emerged with the establishment of a "Steering Committee" to aid in the organization of future meetings. The leader of this committee was Thomas B. Fitzpatrick. I served on this committee that, unfortunately, did very little for the next six years. During this time, a strong movement among the non-melanoma elements of the pigment cell community began to assert itself. This group wished to form a pigment cell society that was truly representative of the community and truly international. Thus, at the IXth IPCC held in Houston in 1975, much discussion led to the formation of a committee consisting of Tom Fitzpatrick, Walt Quevedo, and me, charged with the responsibility of establishing an international society. Later, Vernon Riley, Aaron Lerner, and Alene Silver were added to the committee. We were all Americans, but since I argued so strongly in behalf of the international community, I became their representative. Through the diligent efforts of Dr. Silver, a draft copy of articles of association and by-laws for the International Pigment Cell Society (IPCS) was prepared within a year, and after the committee approved, the IPCS was legally incorporated on March 21, 1977. The implementation of the IPCS was associated with the Xth IPCC in Cambridge, Massachusetts, in October of that year. The second

IPCC to be held under the auspices of the IPCS, the XIth IPCC, took place in Sendai, Japan, in 1980 under the chairmanship of Makoto Seiji. At a council meeting in Sendai, I was asked to organize the XIIth IPCC in Tucson. I declined because I was just not ready for such an undertaking. I was still young, and I knew that from my position in a biology department in a liberal arts college, raising sufficient funds would be difficult. Moreover, my friend Fritz Anders of the University of Giessen was approaching retirement age, and thus it would be more appropriate for him to chair the XIIth IPCC. I argued strongly for Fritz to be named chairman of that conference by pointing out that, if the IPCS was to be truly international, the XIIth IPCC should be held outside the United States. It was just too soon for us to host another IPCC. My argument was convincing, and thus Giessen was chosen for the XIIth IPCC in 1983. The XIIIth IPCC would fall to me in Tucson in 1986.

From my return from the Sendai IPCC, and for the next few years, much occurred that would affect the development and growth of the IPCS and my role in those changes. Actually, the first event began a few months before the Sendai meeting. By that time, I had changed departments due to a very unfavorable environment that arose in the Department of Cell and Developmental Biology where I had been situated. Following the fraudulent use of my stores account by the department head as I mentioned earlier, much turmoil developed in the department. I thought it best to ask our dean to move me to the Department of General Biology. This was entirely an administrative move for me and my students. We occupied the same space, and my teaching roles were unchanged. I knew that this might be only a temporary change since the "handwriting on the wall" indicated that a wholesale reorganization of the biological sciences was in the offing. I was at home in the Department of General Biology whose members were primarily teachers of the majority of undergraduate biology courses. Soon after I joined the department, the head, Ivan Lytle, died unexpectedly. At a departmental meeting at which I was not in attendance, the faculty drafted me as their head. I was put on the spot. I really was too much involved in my own research and teaching efforts to consider the overture; moreover, my international involvements in the worlds of pigment cell biology and comparative endocrinology took some time. However, I felt a strong compassion and concern for my colleagues in the department, for they were very much maligned by the provost and higher administration that viewed them as merely teachers. So, in a quixotic gesture and with the support of the Dean of Liberal Arts, I took on the headship of General Biology for three years. I fought hard for the department, jousting with windmill after windmill, knowing that I had the respect of my faculty and staff. Finally, as I had expected, in 1983 a total reorganization of Life Sciences began. In the process, the Department of General Biology met its demise. The question of where I would go was answered quickly in a phone call I received from Lou Kettel, Dean of the College of Medicine. He asked if I would consider joining the Department of Anatomy in the College of Medicine. The move would be entirely administrative, for I was to remain on main campus in my own office and lab, and my undergraduate teaching would be the same. Dean Kettel's invitation was apparently spurred on by Anatomy's lack of full professors to serve on important committees. There were several permanent associate professors who were not eligible for promotion. My name came up since I had served on several evaluation committees in previous years when the department was reviewed. I accepted the offer, and, administratively, I remained in the College of Medicine until my retirement in 1992.

I relished the loss of my administrative load, for now I was becoming more involved in plans for the IPCC that I was to host in three years or so. Moreover, I knew that I could more effectively organize the meeting and secure supporting funds from a position in the College of Medicine than I could from Liberal Arts. Thus, when I attended the XIIth IPCC in Giessen, I was much more confident about hosting the next IPCC in Tucson. At the IPCS council meeting that was held there, we discussed the exact dates and location for the XIIIth IPCC. In my mind, there was no question about the location; however, during the three years since the Sendai meeting, Aaron Lerner and his group at Yale had decided that they would like to hold the XIIIth meeting in New Haven. Aaron was fully aware of what had transpired at the Sendai council meeting, as were the other council members; however, he argued that New Haven was much more culturally significant than Tucson. I was not about to discuss the cultural comparisons between New Haven and Tucson, but I did point out that we would be ready for the meeting in Tucson three years hence. The Japanese were ready to go to Tucson, and so were most of the Europeans. Moreover, there was a bit of an unspoken resentment toward the domination that the eastern corridor had displayed in organizational matters pertaining to the various IPCCs. A vote was taken, and the IPCS council accepted my invitation to meet in Tucson in October of 1986.

Prince Masahito at his poster presentation at the XIIIth IPCC in Tucson in 1986.

Thanks to the help of many people who helped in many ways, the meeting was a grand success, so much so that I received compliment after compliment, even from some members of Aaron Lerner's group who felt that both scientifically and from the hospitality provided, the Tucson IPCC had been the best so far. I had the last laugh from the cultural standpoint for, by coincidence, Isaac Stern was giving a concert in Tucson during one of the evenings of the conference. For the Japanese participants, one of the high points of the XIIIth IPCC was the presence of Prince Masahito. I had asked Masahito to serve as Honorary Chairman of the conference. I knew from my Japanese friends that, just prior to our meeting, he was scheduled to make a state visit to Brazil where the Japanese population exceeds that of any other place in the world outside of Japan. His office kindly responded to my invitation by stating that the Prince was grateful for the invitation, but that he could not accept at that moment. I was overjoyed, for I knew that he would come after all matters of security were investigated and approved. In retrospect, it is not the science, the presence of Prince Masahito, the visit to the Desert Museum, or the party at Old Tucson that the XIIIth IPCC will be remembered for. Rather, it was the initiation of events that led to a total reorganization of the IPCS, as it was then known.

By the time we gathered in Giessen for the XIIth IPCC, other pigment cell groups, led primarily by the missionary efforts of Peppe Prota in Europe, were developing parallel with the IPCS. I knew that Peppe was interested in starting a European pigment cell organization that would eventually rival the IPCS and that would provide a venue not dominated by the Americans' influence centered in Cambridge and New Haven. He led the establishment of a series of European Workshops on Melanin Pigmentation. The Vth such workshop, held in Marseille in 1984, was important. There and at the VIth workshop held in Murcia, Spain, the next year, seeds were sown for the establishment of the European Society for Pigment Cell Research (ESPCR). Inauguration of this new society, with Peppe Prota as president, was in Naples at the end of 1985. Within one year, the ESPCR had enrolled 120 members, and its council began to assess its relationship with the IPCS. At the same time in Japan, the well solidified Japan Pigment Cell Club began to consider establishing a more formal organization. I applauded the formation of the ESPCR, became a member, and served as an executive council member in 1986. However, deep down I had certain reservations, because I knew that Peppe, my dear friend, was going a little overboard. I was particularly concerned because I knew that he was hell bent to start an ESPCR-sponsored journal devoted to pigment cell research. I felt that if such a journal were to be formed it should be affiliated primarily with IPCS and not with a more regional organization. Moreover, I understood that their journal was to be called *Melanoma Research*, a title that would not be representative of the broad spectrum of pigmentation researchers.

In light of what was going on in Europe and Japan, I realized that the status of the IPCS was going to need a major change for, as it stood, it was still more American than it was international. From the European viewpoint, the IPCS was a *de facto* regional meeting from which they were essentially excluded except for paying dues, and so forth. I felt that it was time too for us in the western hemisphere to establish our own regional society and to join with the ESPCR and the newly forming Japan Society for Pigment Cell Research (JSPCR) in the form of a federation that would replace the

IPCS. This thought had been percolating in my mind for a year or two before the Tucson IPCC. These thoughts were further reinforced when I began to make plans for publishing the Proceedings of the XIIIth IPCC. In the past, this task was always the responsibility of the IPCC chairman. It had in fact become a bit of a problem, especially under the tenure of Vernon Riley who somehow managed to contract for very expensive and poorly done volumes. Because of my earlier five-year tenure as managing editor of the *American Zoologist*, I had developed good relations with our publisher, Alan R. Liss, Inc. Naturally, I turned to them as a possible publisher for our symposium volume. They offered a favorable contract, which I signed. During the year or more of discussions with Liss, I had frequent conversations with Paulette Cohen, one of their vice-presidents, who urged me to start a scientific journal to support our society and discipline. I rejected the idea, thinking it was not my place to do so, and that it should be an action of the IPCS council. The more I thought about it, the more I realized that the inefficiency of the IPCS would preclude its ever happening. It would happen, as the Italians say, "*a ogno morte di Papa*" (whenever a pope dies). I sensed that, if I did not take action, the ESPCR-sponsored *Melanoma Research* would become our pigmentation journal by default. Thus, I undertook negotiations with Paulette Cohen after first corresponding with leaders of the pigment cell community, including Peppe Prota. In fact, it was he who suggested the title *Pigment Cell Research*. The ESPCR had already agreed to act as co-sponsor (with IPCS), and thus *Pigment Cell Research* was launched.

At the IPCS council meeting in Tucson, I rose to champion the view already expressed earlier in a newsletter from the ESPCR, implying that in light of the newly established ESPCR, the forthcoming establishment of the JSPCR, and the potential formation of an American society, the role of the IPCS needed to be reassessed. This point was brought up at the general assembly meeting of the IPCS. The assembly voted to appoint a committee to look into a revision of the IPCS bylaws so that they could accommodate the inclusion of the emerging regional societies. Another committee was to be directed to consider the organization of an "American" regional society. Months before the Tucson IPCC, I had thought about doing the latter on my own; however, I thought better of it realizing that it would be viewed as a power grab. Instead, I made suggestions to others to make such a move.

Eventually, Jim Nordlund from Cincinnati and Dick King from Minnesota organized what would become the Pan American Society for Pigment Cell Research (PASPCR). Its bylaws were approved at the first PASPCR meeting held in Minneapolis in June 1988. At that meeting, the concept of the federation was discussed. An ad hoc committee chaired by Hans Rorsman, IPCS president, decided that representatives of the three regional societies and I, as PCR editor, meet in Lund, Sweden, on October 1, 1988, to get the federation going. A bylaws committee was appointed to present the bylaws for the IPCS council to approve at the council meeting to be held the following September in Venice. The general assembly of the IPCS ratified these bylaws at the Kobe IPCC in 1990, and thus the International Federation of Pigment Cell Societies (IFPCS) was born.

These organizational events that burst forth at the Tucson meeting were generally met with approval, but not everyone was satisfied, especially some of the old guard. At a final plenary session, my respected friend, Tom Fitzpatrick, uttered these words: "At the 1975 IPCC in Houston, Joe Bagnara was instrumental in getting the IPCS started, and here in 1986, he is instrumental in bringing it down." This was not a permanent pique on Tom's part, for we remained friends until his death a few years ago.

The year 1986 was also a time of great personal stress, but with perseverance and understanding, sanity was restored, and the good life continued.

With the termination of the XIIIth IPCC and the editing of the Proceedings volume for the meeting, it would seem that I could find time for rest and relaxation. That was hardly the case, for now I had the responsibility of a new journal on my shoulders. At the outset, things went smoothly enough. There were some good manuscripts available, several from participants of the IPCC. I established a routine for handling the manuscripts, getting them through the review process, and eventually sending them to the publisher. At the outset, I was allowed a relatively small number of pages for each issue, and this made it hard to get a good balance of subjects. Some of the papers needed to be published in a timely manner and others had less urgency. We covered such a range of areas pertinent to pigment cell biology that not everyone was happy with the content of each issue. Nevertheless, things went along for the first two years. The publishing business was pretty tough at that time, and the future of printed journals was becoming a bit shaky. Liss dropped out; Wiley became our publisher; and there was a good possibility that *PCR* was not going to survive since it was not a moneymaker.

In 1990, a wonderful thing happened. Munksgaard Ltd., an established and much respected Danish publisher, bought several journals from Wiley, and among them was *PCR*. In short order, a senior vice-president from Munksgaard, Peter Hartmann, came to Tucson to discuss the journal with me. Dealing with Peter was like a breath of fresh air, and his and Munksgaard's understanding allowed the journal to survive, and now, twenty-two years later, *PCR* continues to exist. Many changes took place that allowed for this success, including a name change and expanded sponsorship from the melanoma community. Thus, the journal is now *Pigment Cell and Melanoma Research*. As founding editor, I continued my role for seven years. Finally, in 1994, as a result of my own introspection, together with considerable pressure from the IFPCS and constructive discussions with Peter Hartmann and Vince Hearing, I stepped down as editor. I have always had the utmost respect for both Peter and Vince, so my decision was an easy one, made for the following reasons. Very important was the fact that, with the establishment of the federation and its sponsorship, the term of the editor was to be for five years. Moreover, the succeeding editor was to come from a different international society on a rotating basis. I was hoping that Vince would succeed me, but, since he was then president of the IFPCS, that was not possible. The new editor needed to come from the JSPCR or ESPCR. Given that molecular genetics was rapidly developing, I felt that we needed someone knowledgeable in this area to be my successor, and I suggested that Taki Takeuchi become the new Editor. Vince agreed and Peter had no objections. As luck would have it, the Federation nominated Taki as the new Editor.

The transition from me to Taki was going along quite well when Taki tragically passed away. Jiro Matsumoto was appointed to finish the remaining term and served from 1996 to 1999. He did a fine job as editor. Jiro was succeeded by Vince, and during his tenure as editor, *PCR* flourished with an increasing citation index.

14

Making a Movie

While much was going on with me and the tadpole patrol throughout the 1970s and 1980s, a continuing preoccupation was my breeding colony of *Pachymedusa*. For most of the year, the greenhouse at my home, where most of the frogs were housed, was an active place. I did have a few pairs in the lab where they were maintained in a large, specially made box that could be kept in summer-like conditions; however, most observations of the behavior of leaf frogs were made at home. Not a day passed when we were in Tucson that did not involve several visits to the greenhouse. During the main breeding months of May, June, July, August, and September, I checked in on them early in the morning just after I got up. If breeding had started overnight, the couples would still be copulating when I looked in on them. I would watch until they finished, and then I observed when they and their colony mates prepared for their daylong sleep in some shady spot in the greenhouse. At dusk, I would again be sure to check the greenhouse to watch as they awakened to begin a nighttime of feeding or, in a few cases, a long breeding session.

As I mentioned earlier, we convinced the frogs in our facilities that it was breeding season by essentially duplicating the summer weather of Sinaloa and southern Sonora. This meant high heat and humidity. The former was no problem thanks to the Arizona sun making the greenhouse more than toasty. In fact, to keep the temperature at a reasonable level, I installed a thermostatically controlled exhaust fan. High humidity was achieved through a timed watering system that delivered a fine spray. Thus, in response to the external cues provided by increasing day length, temperature, and humidity, the frogs were deceived into preparing for breeding and then actually doing so. In females, the former involved vitellogenesis, a process effected by increasing estrogenic hormones from the ovaries. These hormones then stimulated the liver to produce a protein, vitellogenin, which was transported by the circulation to the ovaries. There, the hundreds of thousands of ovocytes took up the vitellogenin for storage as yolk platelets. Induction of breeding for males meant that androgens from the testis induced appropriate male secondary sexual characteristics. These included enlargement of vocal sacs and forelegs and development of a heavy, darkened thumb. It took some time for these preparations to take place, but by the time the high heat and humidity reached levels equivalent of those found in Mexico during the summer monsoons, breeding behavior commenced and coupling took place.

When a particular female was ready to deposit her eggs, she went to a water source, a large plastic tub embedded in the floor of the greenhouse beneath overhanging branches. She carefully lowered her posterior end into the water as she clung to the side of the tub. She remained in this position for about forty-five minutes. A male or males might, or might not, be clinging to her during this time when she was obviously taking up water. When the time was up, she abruptly climbed to an overhanging limb with at least one male aboard. She sought an appropriate overhanging smallbranch or group of leaves and soon began to extrude a few eggs at a time onto the leaves where they immediately stuck. At the same time, the male began to discharge a few drops at a time of essentially

The "stars" of the movie, in amplexus in the greenhuse.

dilute urine containing countless spermatozoa. This fluid spread over the back of the pendant female and over the developing egg mass. This process continued for about fifteen minutes when the female descended back to the water's edge in order once more to take up a volume of water to hydrate the next group of eggs. When she was ready, she ascended to her previously deposited aliquot of eggs, and another group of eggs was released at the top of the egg mass. Once again, the eggs were irrigated by urinary fluid from the male. This process was repeated again and again until a longish mass of eggs was suspended over the water. By the time she had finished, the egg mass may have contained as many as 2000 eggs. What I have described for the greenhouse is essentially what happens during the reproductive period of *Pachymedusa* in Mexico. During the heavy rains, pools of water may accumulate in low spots, and breeding pairs will choose branches over such pools to oviposit. During the next few days or more, the embryos will develop into small larvae in the *in situ* egg mass. When these larvae, or tadpoles, reach the hatching stage, they emerge, flip off into space, and splash into the pool below. Here, they are on their own to try to live through the next stages of their lives.

Somehow, egg-laying, whether it takes place in a greenhouse or in nature, occurs when the female perceives that the humidity is high enough to keep the egg mass from drying out during the week or so before hatching takes place. How she detects the humidity is, of course, a mystery. As I watched and photographed many egg masses over a breeding season or two, I could not help but be impressed with how the egg mass changes over the period that it is suspended over the pool. When the female finally finishes depositing her eggs and leaves the egg mass for good, it is a large gelatinous glob with the eggs well embedded in the jelly just beneath the surface. Each egg is enclosed by a fertilization (or vitelline) membrane that tightly embraces the egg cell membrane. Once the egg is fertilized, the vitelline membrane lifts off the surface of the egg creating a space, the perivitelline space. As the week goes by, the space grows larger such that each egg floats in its own little "perivitelline sea." At the same time, the jelly seems to "melt away" as each egg in its own perivitelline sea projects outward so that the entire mass ultimately looks like a bunch of grapes. I found this very intriguing as I developed the impression that water stored in the jelly passed through the vitelline membrane to provide the fluid that created the perivitelline sea.

I told one of my handball-playing friends, Dr. Allen Cohen, about this observation, and he became equally as excited. Al was a biochemist for the US Department of Agriculture Bee Laboratory located at the University farm, near one of my greenhouses. His curiosity about this observation of water transfer from one compartment to another rivaled my own, so we decided to investigate the problem together. He had the equipment to measure the micro quantities of ions that were undoubtedly involved. As a first step, we measured the total weight of an egg mass just after it was laid, and then on subsequent days until just before hatching. We found that the weight was unchanged from day to day, and thus concluded that during this period, water moved from one compartment, the jelly, to another, the perivitelline space. Before we could do more than a few preliminary measurements on the egg fluids, Al was transferred away from Tucson and further experimentation ended. Thus, this curious and interesting story remains unexplained and untold. It is yet another entry to the list of unfinished business that dots my scientific career.

The question of water retention as it impacts many aspects of the leaf frog's daily life and development is fascinating. In fact, this problem is a basic one that all amphibians face. It is particularly important to the leaf frogs, because at times they are subjected to periods of high heat and low humidity during both the annual and daily cycles. *Pachymedusa*, like other leaf frogs, are primarily nocturnal. During the day, when it becomes exceedingly hot, the frogs rest on a flat surface, such as a large leaf, and position themselves so as to expose the smallest amount of their surface to the dry ambiance. Their arms and legs are held close to the body, their eyes are closed, and their entire ventral surface is plastered against the substrate to form a tight seal. The entire exposed surface becomes quite shiny, as though it has been coated with a layer of lacquer. It is known that one of the South American leaf frogs, in fact, does coat itself with an impervious waxy coat to keep from drying out. This is not the case with our *Pachymedusa*. I have often observed at dusk that, when the frog wakes up and is preparing for its nocturnal period, it begins to wiggle in a peculiar way, much as a human would when removing a wet suit. As the wiggling persists, a thin layer of what looks like skin slides forward and is actually swallowed, little by little. We have not studied this process and, as far as I know, no one else has either. On the basis of these empirical observations, however, it seems that during the hot day the frog develops a thin impervious skin that keeps it from drying out. At night, when the frogs immerse themselves in water or blot themselves on wet surfaces (which they do) to take up water, the membrane must be removed. Since it probably has nutritional value, why not eat it?

These few observations that I relate here are only a few of the many remarkable events in the daily life of *Pachymedusa* that I observed. It seemed to me that the best way to record all these events and their various nuances would be to put them on film. To this end, I again called upon one of my friends from the handball-playing community, Alex Kerstich. Alex was a scuba diver of some renown who taught marine biology at one of the local high schools. He had collaborated with one of my departmental colleagues, and the two had published a noteworthy volume on the fishes of the Gulf of California. Alex did the photography and had taken extensive underwater footage. When I told him about my frogs and my desire to record their life story on film, Alex agreed to partner with me.

So, when my early morning activity as a frog voyeur discovered a pair of frogs *in flagrente delicto*, I would call Alex, often getting him out of bed, to come to my house to film. Thus we obtained some excellent footage of all aspects of *Pachymedusa* biology. Some of the views of oviposition were remarkable. Eventually, after having awakened Alex or his wife in the early morning too many times, Alex left the camera with me, and I added to our accumulated footage. When it became obvious that we had recorded material that was not only unique, but also pretty good, I took advantage of the filmmaking program at the University of Arizona to have our film produced. They liked what we had, so I wrote a script, chose some music, and sought a narrator. He was David Graham, a former undergraduate student of mine who was then a radio announcer. Since most of the footage we selected focused on the reproductive biology of *Pachymedusa*, we chose the conservative title of "The Mexican Leaf Frog and Its Reproduction," although among Alex and his wife and Lou and I, we came up with some less bland titles. The best of these included "I Am Curious Green" and "Deep Croak." In any case, the film came out in 1979 and was fifteen minutes long. I thought it was good enough to sell, so I submitted it to several universities for distribution. The University of California Film Center accepted it for inclusion in their list of titles, and thus it became available for rental for several years.

15

Italy Revisited

Although the Mexican leaf frog film was not available for commercial distribution until 1979, the University of Arizona version was mine. I used it in several presentations, mainly at comparative endocrinology meetings in Japan, Europe, and here at home. When these conferences were held in Europe, my Neapolitan friends, including Giovanni Chieffi and his people, usually attended. A part of that group worked on amphibian reproductive endocrinology and was well aware of the possibilities that the Mexican leaf frog offered in that area.

I had become particularly close to two of Giovanni's associates, Rakesh Rastogi and Luisa Iela, who were members of the zoology department of the University of Naples. Rakesh was a native of India where he had his formal training. Giovanni met Rakesh in Japan and was impressed enough to add him to his group, and in Italy he met Luisa, a born and bred Neapolitan and an *allieva* (student) of Giovanni. They married and became parents of Silvia, a little girl whose mixed heritage produced remarkable beauty. Rakesh was gifted in languages and became fluent in Italian very quickly. In fact, in short order Rakesh became as Neapolitan as a native born *napoletano*. He was so successful that he soon became Professor. Luisa chose not to pursue a rank higher than the Italian equivalent of Associate Professor. This proved to be a blessing in the long run, for when Rakesh and Luisa decided to come to my lab for a year, Luisa was eligible for some funding from which Rakesh was excluded. Thus, Luisa applied for a stipend from the Population Council of the Rockefeller Foundation to work on *Pachymedusa* reproduction. She received the grant, and thus Rakesh, Luisa, and six-year-old Silvia were able to spend about a year (1983–1984) in Tucson. We were grateful to the late Sheldon Segal, Director of the Population Council for this possibility. Shelly Segal had been a PhD student of Emil Witschi and had come back to the lab in Iowa City when I was finishing my graduate work, and Giovanni and Annamaria were spending their year there. The arrival of Rakesh, Luisa, and Silvia in Tucson was the beginning of a long collaboration and the development of a profound, lasting friendship.

Rakesh and Luisa

It took little time for the Rastogi family to become embedded in Tucson life. It helped that Lou and I already knew and liked them. We helped them find an apartment near the university farm and just over two miles from our house. They soon bought a cute orange VW beetle and they were in business. Silvia was enrolled in a public school nearby. As a good Neapolitan housewife, Luisa needed a good, healthy basil plant, and for this I was happy to oblige. I have long been a grower of basil plants, and in those days I had a supply of different varieties. I enjoyed giving basil plants as gifts, because I had learned from Greek friends that to do so is to bring good luck. This is why in Greece one sees a basil plant outside of every door and at the foot of every bus driver.

At the time that Rakesh and Luisa arrived, the lab was a busy place as we worked on a variety of projects. I was really grateful to have competent hands to take over on the *Pachymedusa* reproductive biology aspects. Although the Rockefeller award was for Luisa, Rakesh really took the lead, as the work became a collaborative effort between my lab and his. A major flurry of activity began when Rakesh and Luisa were in Tucson, but over the next ten years or more our collaborative efforts switched to Naples where some of Rakesh's graduate students took part. A series of solid publications resulted under the same umbrella title, "Reproduction in the Mexican Leaf Frog, *Pachymedusa dacnicolor*." The first paper in the series was a monograph mostly based upon observations that I made in the 1970s. It was entitled "I. Behavioral and Morphological Aspects." Papers II and III were mostly the work of Rakesh and Luisa and were entitled, respectively, "II. The male" and "III. The female." Paper "IV. Spermatogenesis: A light and ultrastructural study" was done by Rakesh and me with the help of Peggy

Krasovich, our electron microscope technician in the anatomy department. One of the interesting findings of this study was that the spermatozoan of *Pachymedusa* was not a simple flagellum, but rather had an undulating membrane attached to it. What this means is a mystery to me, but I hope that someday someone might take up this problem and look at it from the standpoint of the evolutionary biology of leaf frogs. If not, mark one more entry to the growing list of unfinished business. The remaining papers in the series were derived from work done in Naples and were concerned with the immunohistochemical localization of gonadotropin-releasing hormones in the *Pachymedusa* brain. Although I was a co-author of these papers, my contribution was miniscule.

Sorrento 1985

The writing of the first three papers of the series took place in 1985. I did much of the first paper in Tucson while Rakesh and Luisa wrote drafts of the second two in Naples. Since I was eligible for a sabbatical, we planned that Lou and I would go to Naples for a few months during the spring semester of 1985. During that period, we would ready the three manuscripts for publication and begin work on the fourth. There was no way that we would live in Naples again, because of all the commotion and noise. Instead, we planned to rent an apartment in Sorrento for the spring before the start of the tourist season. The Sorrentine peninsula is quiet, beautiful, and in easy access to friends and work in Naples. A commuter train line, the Circumvesuviana, connects Sorrento and Naples, and it is especially convenient to the University of Naples. The Neapolitan terminus (*capo linea*) is not much more than a ten-minute walk to the building where the zoology department is located. We were to have a car in Italy, but the Circumvesuviana was much more convenient than driving to and from Naples.

In mid-February of 1985, we left for Munich in order to pick up a new BMW 318 that we had ordered. We had already arranged for an apartment, thanks to Rakesh who had scouted out availability in Sorrento. There is a fairly good supply of furnished apartments for rent in early spring because many are owned by people who use them only during the summer, and they are thus available during the off-season. The apartment he chose was conveniently located in a newish complex of apartment houses in the middle of town and came with parking in the basement. It was still winter in southern Germany, and as we headed south, we encountered snow. I can still remember snow crunching under our feet as we took an evening stroll in the village where we stayed our first night after leaving Munich. We crossed the Brenner Pass into Italy and spent the next night in Modena, Luciano Pavarotti's hometown, where we found a hotel that had guarded parking. The people of BMW made us quite conscious about theft in Italy, and they urged us to purchase wheel locks before we left Munich. As we drove south, the urgency to install them grew, so before we found lodging in Cassino at our next stop, I found a wide place along the Autostrada to install them. I did so, although I felt that it was much ado about nothing and that perhaps the admonition of the Germans was a bit of an overreaction. Nevertheless, as the Italians say, "*la prudenza non è mai troppa*" (you can't be too careful). Fortunately, no one ever stole our wheels. Off we went to our *rendezvous* with Rakesh, Luisa, and Silvia, who led us to the red-shuttered apartment building in Parco Snitzer.

For the most part, the apartment was comfortable. It was on the fifth floor and had two ample terraces. There was an elevator, but we often took the stairs for exercise. Unfortunately, there was one major problem. It was exceedingly cold, and in fact, it was downright frigid. The policy in many Sorrentine apartment buildings was that there was no central heating during winter except for vacation periods. After all, spring would come, and it would soon warm up. Unfortunately, February and March of 1985 were unusually cold and rainy. The former made it uncomfortable and the latter made the drying of our laundry interminable! Thanks to the thoughtfulness of Rakesh and Luisa, who lent us room heaters, we were able to make do, and we didn't freeze as I worked on manuscripts or as we read or watched Italian television.

As spring approached, we were better able to take advantage of the beauty and the location of the Sorrentine peninsula. The peninsula projects outward to the southwest from what would be the shin of the boot of Italy. In doing so, it partially encloses and cups one side of the Bay of Naples. Just off the tip of the peninsula, and separated from it by a narrow strait, is the island of Capri. Judging from geological and biological criteria, the two were connected in ancient times. Sorrento is located on the northwest side of the peninsula, not many kilometers from the tip. The major road connecting the autostrade and Naples to Sorrento essentially goes along the bay and

A view of the east side of the Bay of Naples showing the contiguous towns of Sorrento, Sant' Agnello, Piano di Sorrento, and Meta di Sorrento that end at the promontory of Montechiaro beyond which another string of towns, including Pompei and Herculaneum, lead past Mt. Vesuvius toward Naples.

in several places it traverses tunnels of some length. About six kilometers from Sorrento, the road passes through Meta di Sorrento, the first of three villages that are sequentially contiguous with Sorrento. At this point, the road is called *Corso Italia*. It continues through Piano di Sorrento and then Sant'Agnello, just two kilometers from Sorrento. Up to this point, the route has been scenic, but as it continues on through the town and bifurcates, the roads become spectacular. Just out of town, at Capo di Sorrento, one route goes straight ahead and onward toward the tip of the peninsula. This road is called *Nastro d'Oro* (gold ribbon) as it goes on through two villages, Massa Lubrense and Termini. The views toward Capri are gorgeous. At the Capo, another route ascends the hills away from the sea. This is *Nastro Verde* (green ribbon), probably so named because it passes through wooded hills to the top of and over the peninsula down to the other side. In doing so, it passes through Sant'Agata sui Due Golfi, so named because Sant'Agata is essentially on a ridge from which the Bay of Naples can be seen on one side and the Gulf of Salerno on the other.

We often enjoyed both roads *en route* to favorite places. One such place, very close to Sorrento, was Bagni Regina Giovanna. This is a point at the very end of the Capo Sorrento where on flat rocks in the bay, the Romans had built a temple and baths. Roman walls still remain as relics of their presence. What I enjoyed most about going there was the descent from the Capo through olive groves to the sea.

When spring was in the full bloom of poppies, oxalis, and wild garlic, what a sight it was to drive to Termini where, after an espresso at the local bar, we hiked the scenic trail to Punta Campanella at the very end of the Sorrentine peninsula. The views are wonderful, offering on one side a view toward the Faraglioni cliffs of Capri, and on the other, the Amalfi coast all the way to Salerno. We always took our time on this hike, because there was so much to see. The vegetation was lush, and there was good nature watching, including lizards by the hundreds sunning themselves on rocks, falcons skimming along the cliffs, and many songbirds. I was lucky enough to spot a black-capped warbler (*capo nero*) nest right along the trail. The trail passed near a substantial *Torre dei Saraceni* (Saracen Tower). The southern Italian coast is dotted with such towers that were built as watch towers to spot invading marauders from North Africa. These invaders had a profound effect on the entire length of the southeastern coast of Italy from Calabria to the Ligurian coast near Genoa. The presence of their genes is evidenced by the swarthy skin of many people along the coast. In places, they formed colonies that took centuries to absorb. One such colony persisted along the beach below Piano di Sorrento where, according to a friend who was born there, Arabic was still spoken when he was a boy. The *Torre dei Saraceni* are substantial structures that have endured. We knew of two along the Amalfi coast that were converted to a hotel in one case and to an elegant residence in the other. According to a friend, it was the residence of Franco Zeffirelli.

While we were busy enough in Sorrento, at least three times a week I took the Circumvesuviana to Naples to write. Many times we would go back on a Saturday or Sunday to be with Rakesh and Luisa, the Chieffis, the Protas, or the Gaetanis who often invited us for *pranzo*. These were fascinating rides not only because of all there was to see as we passed famous places, but to observe our fellow passengers. Usually, we went in the morning in the company of commuters or students whose destinations were jobs or schools in towns along the way. Castellammare di Stabia, the largest city *en route*, is an important port and dry dock location. Here, many descended as new riders boarded. A few stops ahead is Pompei Scavi where there were always a few tourists, Brits and Germans mostly, ready to disembark to spend the day taking in the sights of ancient Pompei. The next significant stop is at Torre Annunziata where, as the train approaches Mount Vesuvius, one can see evidences of lava flows on its slopes. Experts can identify particular lava flows with the year the eruption occurred. Just past Torre del Greco, another important small city, the train arrives at Ercolano (Herculaneum), the site of mass inundation and death during the famous eruption of 79 AD. The tourists who descend here to walk down to the *scavi* are provided with a fascinating experience. The train soon arrives in Naples where it empties as everyone else leaves for the day. Many are students and faculty at the University of Naples.

We knew our way to via Mezzocanone and the university where we met Rakesh and Luisa and the students. We discussed our manuscripts, or perhaps I was to give a lecture. Sometimes I interacted with Giovanni Chieffi or with Peppe Prota. As noon approached, we generally headed out for a pizza in this neighborhood near the Spaccanapoli. For certain, afterwards we would go to Bar Nilo for espresso. It was so named because it is situated in the Piazzetta Nilo where there is a Roman sculpture representing the River Nile. A few years later, when I became a contract professor at the university, all the workers in the bar knew me. I always had a double espresso (*doppio*) and so, when they saw us coming, they put the *doppio* on to brew for me to have immediately on arrival. The ride back to Sorrento was considerably less hectic, because we chose to return in mid-afternoon. The vistas were very beautiful as we looked out on the bay with Capri and Ischia in the background.

We did an enormous amount of walking in Sorrento. There are many small streets and routes to the edge of the cliff on which Sorrento is located. We could look across the bay to Naples or Vesuvius or Castellammare at the base of the 1100 meter Monte Faito. Sorrento is the lemon capital of Europe and, despite the loss of many lemon groves to development, there are still many in production. Many individual groves are separated by small ways or paths defined by ancient stone walls encrusted with mosses, lichens, liverworts, and small flowering plants. They are beautiful, and we loved to walk these paths. Protecting the lemon trees are scaffoldings and frameworks upon which reeds or plastic are placed as protection from the elements. The poles used for these frameworks are actually chestnut saplings. As I learned in later years as I hiked the hills and mountains of the Sorrentine peninsula, the slopes are heavily forested with edible chestnut trees. Most of the large trees were cut down many, many years ago, and from their stumps a steady supply of saplings grow. When they achieve an appropriate size, these are harvested to insure a continuing supply of posts.

From time to time on Sundays, we would take the *aliscafo* (hydrofoil) to Capri to spend some time with Marilyn and Gabriel or to take a hike. We would walk to their villa which, by now, we knew quite well. Later we would walk to a favorite restaurant on Capri, or take a local bus to climb to Anacapri where we went to a small restaurant somewhat hidden from the usual tourists. There are some beautiful places to walk on Capri that are away from the usual daytime hordes of tourists. One of our favorite walks leaves the piazetta through a small commercial alley from which a footpath leads on to the east end of the island and to spectacular views of a beautiful natural arch (*arco naturale*). This point affords a great view of Sorrento, the small towns on the *nastro d'oro*, and *Punta Campanella*. Backtracking a bit takes one past the start of a steep descent by hundreds of steps to a path just above the water. The path is heavily vegetated at this point and passes by a shallow cave, Grotta di Matromania, with a ceiling blackened by campfires that are said to have been made by the Romans. The path continues on just above the Faraglioni cliffs, immense boulders that rise up from the surface of the sea, to offer a gorgeous sight. The path then continues upward, with the aid of steps in places, until it arrives at Belvedere di Tragara, a viewpoint with a fabulous view of the Faraglioni. The route back to town on a wide, well maintained promenade takes about twenty minutes and passes elegant hotels, shops, and boutiques.

While most of our sightseeing during this 1985 stay was local, taking in the wonderful offerings of the Sorrentine peninsula, we did take a short trip that took us away from our cold apartment for a few days. We had both read about the interesting things to see in Puglia, the region that makes up the heel of the boot of Italy. Since it is warm in Puglia and reasonably close to our region, Compania, and easily accessible by a relatively new autostrada that transverses the country, we decided to make the drive. Our new BMW mostly sat in the garage under our palazzo, so it was time to take it on the road again. Off we went, and it wasn't long before we drove through fields newly green with the blessings of springtime. After stops at a few touristic sites, we arrived at our first destination, Matera, a small city of about 60,000 that is situated in the region of Basilicata. Among the features of Matera are extensive cave dwellings that date back to the Stone Age. These caves known as *Sassi* were inhabited in some cases for 9000 years. The town of Matera was founded by the Romans in the third century BC. We were looking forward to seeing these *Sassi* where people actually lived until 1950.

First, we had another matter to attend to. Before we arrived in Matera, we developed a problem with the window on the passenger side of our BMW 318. We could not get it to raise, and it remained open permanently. We consulted one of the booklets that came with the car and found that there was a BMW garage right there in Matera. We found it and had good fortune in that they had a replacement for the defective part and, with a certain amount of pleading, were able to convince the mechanics to repair the window right then and there! It was amazing that here in the relative boondocks was a BMW garage that was fully stocked with parts! Actually, we were not totally surprised for, a few weeks earlier, when it was time to have a required service on our new car, we took it to a remarkable garage in Salerno. We had made an appointment for our service and, when we arrived, it was like going into a hospital. Everything was as clean as a whistle, and the mechanics were all dressed in pristine white smocks. The car was placed in its own stall surrounded by white drapes! In any case, back to Matera, the window repair went quickly, and we were able to go to the *Sassi* to complete our Matera visit.

Alberobello in Puglia was our next destination where we looked forward to seeing the famous *trulli*. We were truly impressed by these fourteenth century buildings. They look like enormous beehives and are constructed from flat limestone rocks. Some *trulli* are used as homes, and others are used for a variety of purposes, including hotels. Our hotel was not a *trullo*, but we did have dinner in a restaurant that was a *trullo*. Some *trulli* seem to be connected to almost form a complex, and I have read some descriptions of these groups as looking like a fairyland. That was not my impression, but what did appear to be a fairyland to me was the agricultural panorama of many small fields circumscribed by stone walls that separate one from another. The stones seem to be the same limestone pieces that were used to construct the *trulli*. Most likely, when the fields were cleared for cultivation, all these large stones or rocks were piled up to form walls. In the early morning following our arrival, we drove through the fertile farmland near Alberobello, and as we drove over a hillside into a valley, the early morning sunlight shining on the bright green, rock-lined fields, was magical. Our time was short, and we needed to return to Naples, but at least we had an introduction to this rich province. We needed to return to absorb the many cultural and historical treasures that are offered by this heel of the Italian boot, but that has not been possible, and thus such a lack is another aspect of unfinished business.

The writing project and scientific interactions went well and efficiently, so as Easter approached toward the beginning of April, we began thinking about moving on. Our scientific sojourn was to be mostly in Naples and Sorrento, but I also had some discussions and seminars scheduled in Paris. The latter began to weigh on my mind, and I thought that we should head north a week or so after Easter. This led to a flurry of social activity. On Good Friday, Rakesh, Luisa, and Silvia came to us in Sorrento with the plan to watch one of the religious processions that each of the towns take pride in. We missed the main one in Sorrento that had taken place in the morning, but after they returned to Naples, Lou and I were able to see one in the evening. It was fascinating. The next day we were invited for *pranzo* in Termini where Peppe Prota and Giovanna had a vacation home. For Easter Sunday, Marilyn and Gabriel had asked us to come to Capri. Monday following Easter is *Pasquetta*, also an important holiday, and we were again asked by Peppe and Giovanna to come to Termini for *cena*. We rested for the next few days and started to pack. On the following Sunday, Rakesh and Luisa had us for *pranzo* in Naples where we returned all that we had borrowed from them and started our sad farewells.

We left a few days later with a first stop in a lovely inn near Siena. We arrived in Nice the next day where we were to spend two nights with dear friends Claude and Pierre Freychet. We had met them in Paris in 1968, and during the ensuing seventeen years we became good friends as we saw them several times, both in Tucson and in France. The last leg of our trip took us through Burgundy as we sampled its famous cuisine and ambiance for two days before arriving in Paris. Again, for the how many-eth time, we basked in the Gallien family hospitality. Suzanne offered us her apartment for the next two weeks as she stayed at her mother's just around the corner.

The Paris stay was a whirlwind of science and social obligations. Claude Gallien and I had much to discuss at his lab at l'Universite Paris Decartes on rue des Saint Pères. I spent a couple of days at the natural history museum on rue Cuvier where I presented some seminars about *Pachymedusa*. Perhaps my most prestigious outing was at Nogent-sur-Marne where I spent the day with Nicole le Douarin and her colleagues, and where I spoke about the neural crest as a source of stem cells. Finally, we dropped off our BMW with a shipper who was to deliver the car to us in Tucson. On May 6, we flew home to Tucson, tired and a little homesick, but delighted with our latest experience in Italy and France, countries that we know and love.

16

Again and Again

The two-month stay of 1985 in Sorrento opened the door to a series of almost annual stays in Sorrento that continued for about twenty years. During these visits, we had the best of both worlds. The Sorrentine peninsula was an ideal place to live, while Naples provided the scholarly interactions that were fundamental to my life and career. The two worlds were separated by a mere hour on the train. These Italian trips were not all dedicated to my activities, but also related to Lou's work as a travel agent.

We returned to Naples in 1986, but only briefly on our way to Sicily. In 1987, we again stayed in Sorrento, but only for a few days as I participated in the inaugural meeting of the ESPCR (European Society for Pigment Cell Research) that Peppe Prota had organized. I had dual roles in participating. First, as editor of *PCR*, I had promised to publish the proceedings of the meeting. Second, I was a charter member of the newly formed society and served on its executive council.

Our next two visits to Sorrento and Naples were formal ones, as I served as a visiting professor at the University of Naples. The first was during the spring of 1988 when I gave a few lectures at the University over a two-week span. Again, we wished to stay on the Sorrentine peninsula and to travel to Naples only on days when I was to lecture. We had a stroke of good luck in finding housing. When Rakesh began searching for an apartment for us, he consulted with Gaetano Ciarcia, one of the young faculty members in the neighboring Department of Physiology. At that time, Gaetano and his wife, Cristina, lived in Sorrento. They knew that a nice, new, furnished apartment would soon be available in Meta di Sorrento and kindly made inquiries about its availability for this American professor. It was not quite ready, but they rented it to us anyway. What a jewel of an apartment it was, on the top floor of a small stone palazzo, right in the center of Meta. It was owned by the Porzio family, Mario and Fernanda and their three children, who occupied the lower two floors. Mario Porzio was a sea captain of an oil tanker owned by a major oil company. He was at sea when we arrived in 1988, so all our dealings were with Fernanda, a lovely, kind lady. Our apartment had a nice terrace replete with potted plants and large glass doors that provided a bright, sunny ambiance. A new and modern kitchen was more than we could ask for. The post office and shops were nearby, and the Circumvesuviana station was just a short walk away. Sorrento was only a six kilometer walk away or two stops on the Circumvesuviana. We were in business!

During those years that I was visiting professor, I got to know Rakesh's students rather well, and I had a fairly good notion of what was going on with their research. As I pointed out, Rakesh's research dealt fundamentally with the endocrinology of amphibian reproduction. The lab members functioned as a congenial and effective team. Every year at the height of the spring breeding season, the group made an outing into the mountains northeast of Naples near the town of Montella. In the area are streams, marshes, and ponds where these amphibians breed. It

Lou, Luisa, and Rakesh on the terrace of our jewel of an apartment in Meta di Sorrento.

is a remarkably beautiful place that Lou and I were both able to see because, on at least two occasions, our stay coincided with this annual rite of spring. The wildflowers, including various species of wild orchids, were amazing. This area is also home to a race of chestnut trees that have been grown there since they were introduced by the Greeks who occupied the province in the fifth or sixth century BC.

On one of our trips, we took picnic lunches, but on another we ate at a rustic restaurant called La Bussola (the compass). I had known the restaurant from other trips with Rakesh and very much enjoyed the traditional food that they served. On this occasion, we all sat at a large table, much like a big family. I believe that most of us had pizza. One of Rakesh's students, Biagio D'Aniello, was in a sense my kindred spirit since we were both amateur naturalists and had in fact made field trips together. At any rate, during the course of the meal, we were all impressed with the volume of food Biagio put away. It was especially impressive since Biagio is not a large person. One of the young women commented about Biagio's appetite, and he responded with the announcement that he had once eaten seven pizzas at a sitting. She asked, "*Normale?*" He replied, "*No, una era super!*" We erupted with laughter as Biagio was then and there anointed with the nickname, "*Sette Pizze*" (seven pizzas).

In 1988, Lou could not stay in Italy as long as I was to be there because April was a busy time in the travel business. The same was true in 1991 when we were again able to rent the Porzio apartment. On this occasion, I served as "*Professore a contratto*" (contract Professor) paid to give ten lectures at the University in Naples. This was a formal appointment that the Italian Ministry of Finance recognized by bestowing me a plastic card bearing my "*codice fisicale*" (tax code number).

I felt very fortunate to be asked to come to Naples to lecture as often as I was, and I'm sure that in large measure I was so graced through the actions of Rakesh. The fact that Giovanni Chieffi and I were good friends probably helped, as did the fact that my textbook, *General Endocrinology*, had been translated into Italian and was widely used. The latter fact was brought home in a rather humorous way. One day, before one of my lectures, as the students were filing in, I overheard a young woman ask one of her classmates as she spotted me, "Who is this guy?" She had assumed that I was a foreigner who wouldn't understand Neapolitan dialect. The response was, "Professore Bagnara" which led to the next question, "He of the book?" My lectures went fairly well, even in my simple Italian, thanks to my use of countless slides. I realized all too well the old adage—a picture is worth a thousand words. The students were very kind and patient and obviously some were paying attention for, when I occasionally used a dialect word, I got a great response. I am sure that they wondered how I knew that the word for hole in dialect was *pertuso* instead of *buco* as in formal Italian.

The lectures ended just before the lunch hour and, after a short break to unwind, Rakesh, I, and whoever else was around often walked up the street to one of the numerous pizzerias for a quick lunch. If Lou had come to town, Rakesh, Luisa, Lou, and I would walk to the nearby Spaccanapoli to have a pizza at De Matteos, by far the best pizza in Naples. In later years, it gained much notoriety when Bill Clinton had lunch there during the Naples G7 meeting. Of course, afterwards we returned by way of Bar Nilo for espresso.

After Lou had returned to Tucson and I was alone in Meta, I was able to keep busy by preparing my lectures and doing a fair amount of walking. I am grateful to Gaetano Ciarcia for introducing me to several places to hike in the hills above Meta and Piano. It is interesting that, while Meta, Piano, and Sant'Agnello are contiguous towns, they are each separated or defined by a steep gorge or canyon that transports water from the hills above. These canyons (*vallone*) are fascinating as I learned from Cristina Ciarcia who taught school in Piano. As a class assignment, her students had taken a survey of what there was to see in the *vallone* separating Meta from Piano. It was called La Vinola. They found that the Romans had been there and left much evidence of their presence. On one of my first hikes with Gaetano, we walked up La Vinola through woods of chestnuts. After a few kilometers, we arrived at a beautiful glade called *Selva della Tomba*. Here we found an historic little monument to countless souls who perished during the Great Plague. At this remote spot, a deep pit had been dug to accommodate the bodies of victims of the plague. It seems that the cadavers were hauled up from the coast on carts, dropped into the pit, and covered with quick lime. I am not sure that many people know about the existence of this historical place, but to me it was a moving experience to see and contemplate. In any case,

La Selva is a beautiful last resting place surrounded by trees and greenery. Gaetano and I continued on up a few more kilometers where the trees gave way to grasses and shrubs as we reached the summit of Monte Commune, a smallish mountain of a bit over 700 meters. The view was beautiful as we looked down on the south side of the peninsula at the Gulf of Salerno and the Amalfi coast to the east. Immediately below us was a group of many small islands, Li Galli, so called because they bring to mind a small group of chickens in a hen yard. This was a good spot to consume the great lunch that Cristina had prepared for us. Monte Commune has a special place in my memory for, among the many small shrubs growing there, was what I knew as santolina, a low gray-green shrub that is used as an ornamental in Tucson. I had not known that it is native to the Mediterranean basin.

There were other hikes that I took with Gaetano, but the most memorable is one I took alone in 1991. I had learned from Fernanda that among the more notable Easter processions in this area was one that takes place on Easter Sunday from a hamlet, Santa Maria del Castello, located in the mountains above Positano on the Amalfi coast. From this site at 685 meters, the procession, bearing a wooden cross from the church of Santa Maria, descends the steep slope to Positano. One beautiful morning when I had nothing pressing, I decided to walk to Santa Maria del Castello. I thought that I would follow paved roads as far as I could and then cut cross country through the fields to Santa Maria. I started upwards to the village of Arola, just above Meta, and continued the ascent to the larger town of Preazzano at 477 meters. The next town, Ticciano, seemed a little out of the way, so I asked farmers in the fields to direct me toward Santa Maria. I'm sure they thought I was nuts, but they pointed me through the fields anyway. Thus far, the walk along the road was interesting enough, but now it became quite beautiful as I walked through the springtime greenery. The flowers were magnificent, and I remember one field vividly for it was one solid mass of white narcissus in full bloom. The array of wildflowers to be seen on the Sorrentine peninsula is staggering, and I wish I knew their names. As I passed through a few moist glades, I did see two flowers I knew. The most common were jack-in-the-pulpits of two varieties and wild cyclamens (*ciclamini*). The latter have a special place in my heart because I remember well my mother telling us about the wild *ciclamini* that abounded near her home as a child. As I wandered forward blindly, I finally arrived at Santa Maria del Castello. Now, I needed to find the way down to Positano, so I thought that I would ask at the bar which I recall was quite close to the church. Here, I had no luck, for no one seemed to know the exact route. They headed me toward a small road going out of town and after that I was on my own. I knew which direction the sea was, so off I went. The way was very steep and heavily terraced with grape vines and little plots of crops of various sorts. I was totally lost, but I knew that I needed to go down. At times, I found farmers' paths, and these offered encouragement, but sometimes they ended abruptly, leaving me among dense growth that I had to push through. These places gave me much concern because I knew that the peninsula was home to vipers, and I was in no mood to be bitten. Finally, I found a continuous open path, and soon I could see the Amalfi drive below. My plan was to take a SITA bus back to Meta, so now my hope was that I would hit the highway somewhere near a bus stop. We had taken the bus to Positano a few times, so I knew that there were two stops on the highway above the town. One was at the west end and the other at the east end, and between the two, the town tumbled down the slope to the sea. Actually, I arrived at a point a bit west of town and not far away there was a restaurant or bar. The kind proprietor told me that the bus stop was just a few meters away and that I could buy a ticket from the driver. A bus arrived in short order, and to my good fortune it was one that passed just in front of the post office in Meta instead of stopping on the Corso above the town. Thus, I returned home in forty-five minutes or so rather than in the four or five hours that the twenty or more kilometers ascent and descent had taken. I was pretty well satisfied with myself, especially since I was not very tired, but I was a lot younger then.

 At times, I needed to go to Sorrento. Sometimes I walked the six kilometers through Piano and Sant'Agnello, but at other times I took one of the local yellow buses. In the mid-morning, there were not many passengers and often among these few were older men, mostly retired mariners. More than likely, they were former ship's officers *en route* to town to socialize and to shoot the breeze with their compatriots. Of course, they spoke in dialect, but I was able to comprehend much of what they said because, when I was a boy in Rochester, I heard a lot of Sicilian and Calabrese dialect. The language of the various southern Italian dialects is somewhat similar. It certainly was fun to listen to the banter of these old mariners. I should point out that there were so many retired sailors in the area because Piano di Sorrento is home to a significant maritime college that produces a steady flow of officers who were mostly from the Sorrentine peninsula. It is only natural that they would return home after retirement. Our own Mario Porzio is an example. He was born and reared in Piano and attended the maritime college.

During three stays in the Porzio apartment in Meta, we developed a warm friendship with the Porzios who were most kind to us. In 1988, we knew only Fernanda and the three children, Luigi (Gigino), Angelo, and Gabriella. In 1991, Lou went home before Mario returned from a tour of duty, but I met him and enjoyed his company. From time to time, I was asked to join them for *cena*. I became friends with Gigino who had just received his degree in pharmacy and apprenticed in various pharmacies in the area. I was extremely grateful to Gigino when, one morning, he was kind enough to get up at an ungodly hour to drive me to Capodichino airport in Naples to catch an early flight. Our last stay in Meta was in 1993, because afterwards the apartment was no longer available since Gigino, now an adult and out on his own, staked claim to it.

Parco delle Petunie

When we returned to Sorrento in subsequent years, we needed a new place to stay. Again, it was Rakesh who paved the way. He contacted an agency in Sant'Agnello through which we were able to rent an apartment that was quite comfortable and convenient. It was a ground floor, three rooms (plus kitchen) apartment with its own garden in an apartment complex, Parco delle Petunie. Between 1995 and 2004 we were able to rent it on five occasions for varying lengths of time. In more recent years, we have stayed for short periods in a small hotel in the neighborhood.

We kept returning to Parco delle Petunie for several reasons. First of all, by renting the same apartment on all five occasions, we became quite familiar with all of its idiosyncrasies. The fact that the kitchen, each bedroom, and the *salotto* (parlor) all opened up to the garden was especially appealing. The location of the Parco was excellent. Its gate opened onto *via Cappuccini*, an ancient street that extended from the main street of Sant'Agnello, Corso Italia, almost to the sea. Like many of the streets that probably date back to medieval times, it was paved with large basalt blocks that had been worn smooth over the centuries. It was a short walk to the Corso to find all that we needed, especially groceries, fruits and vegetables, and bread. A side street led to the main *piazzetta* of Sant'Agnello and the Circumvesuviana Station.

Following another sequence of side streets took one to a small family *trattoria* that featured superb pizzas. It was called Moonlight, and to accentuate its name, over the bar hung an illustration that featured two boys with their drawers down and in a position to "moon" the customers. Over the years of our frequenting Moonlight, we became very friendly with the owner and his young sons who were usually our servers. All of our American visitors were taken to Moonlight for dinner for a reasonably priced, excellent meal. We remember very well one evening when, with our dear Tucson friends, Paul Maseman and Lucinda Davis, we went to Moonlight for supper. Lucinda is a vegetarian and, while her husband Paul is fond of meat, he caters to Lucinda's dietary preferences. This evening, however, after countless meatless meals, Paul craved meat. We had never noticed any meat dishes on the menu; in fact, we rarely ever used the menu. I told Pietro, the chef and owner, that our friend would like meat, and he replied, "*Non t'preoccupe*" (don't worry). When the meals were brought to the table, there was a heaping platter of grilled meats for Paul. There was so much on the platter that Lou and I feasted as well.

Among other joys of Sorrento that we shared with Paul and Lucinda, and with others who visited us in Parco delle Petunie, was Marina del Cantone and the small village it serves, Nerano. This village and beach are situated on the Bay of Salerno side of the tip of the Sorrentine peninsula. The beach site, Marina del Cantone, was discovered by the Roman emperor, Tiberius Nerone, who spent much time on Capri where he built his famous villa. Apparently he came upon the Marina del Cantone on a boat excursion from Capri. He was so enchanted by its setting and beauty that he built another villa on the slope above the beach and named it Neronianum. This site became the village of Nerano.

During our various stays on the Sorrentine peninsula, Rakesh and Luisa often came to visit us, and we explored the Nastro d'Oro and Nastro Verde together. We frequently dined at Cantuccio, a restaurant at Marina del Cantone. The restaurant, supported by pilings, extends out over the water.

We reached the restaurant by walking across a stony beach to the shore side entrance. Another entrance was on the water and was accessible by boat. When we first started to frequent Cantuccio, our hostess and server, and presumably a member of the owning family, was an attractive woman in her early thirties. Her dark eyes and skin probably signaled a significant North African contribution to her gene pool, thanks to Saracen marauders. After repeated visits there over the years, she recognizes us and greets us with an embrace and kisses to both cheeks. Like most restaurants in Marina del Cantone, the featured fare is seafood; however, their typical Sorrentine cuisine is what I love. Instead of a grilled fish, I always prefer a mixture of *antipasti* and *contorni* to be accompanied by a simple white wine from the peninsula. Almost all of our visitors have experienced Marina del Cantone and Cantuccio, and most of them, like Paul, consider it their favorite place. I believe that part of the pleasureful experience is the getting there. Most often we have driven from Sorrento, and it is a magnificent drive. We and some of our visitors also enjoy getting there by SITA bus. It leaves from the Circumvesuviana Station and, after numerous stops, it terminates at Marina del Cantone. One of the features of the bus ride is its cross section of passengers from school kids to local residents to tourists.

After having discovered Marina del Cantone and Nerano, Lou and I revisited the area to explore a little. From what must be the main square of Nerano, a wide spot in the serpentine road that descends to the marina, a path departs to the south. I was curious about this trail, in particular because we once saw donkeys at its junction with the piazza. They seemed to be a working pair since they were saddled with canisters that I assume were used to transport materials to and from the lower slopes of the peninsula. More than likely, at some time during the year they were used to haul olives from the terraces that abound in olive trees. We walked down the path that affords a gorgeous view through the olive trees to the blue sea below. After a short distance, we passed the entrance to Villa Rosa. Later we learned that the famous English writer, Norman Douglas, stayed there. Farther along, we came upon a lovely little sanctuary that had been built into the limestone slope. There were potted plants in full bloom gracing this serene place. Obviously, nearby residents maintain the sanctuary and water its plants. Often we have found elderly neighbors reposing in the cool shade. After our first visit there, Lou and I felt that the path must serve the needs of farmers along the peninsula and that it might eventually lead to the tip. With this thought in mind, we made a subsequent visit carrying a picnic lunch. A bit past the sanctuary, the terrain leveled off and the land was cultivated. We soon encountered an elderly lady approaching us on the path. She was bent forward as she carried a load of hay on her shoulders. By this time, we could see the end of the peninsula where there stood a magnificent tower. It was the sixteenth century tower, Torre di Montalto. We made our way up a little slope where we ate our lunch at its base and took in the magnificent view. To the southeast, we could look past the marina below us toward Positano and the Amalfi coast. To the west, we could see past Punta Campanella at the end of the Sorrentine peninsula, to the Faraglioni cliffs of Capri. I was so taken by this site that, a few years later, I led Rakesh and Gaetano Ciarcia there for a similar picnic.

During one of our sojourns at Parco delle Petunie, Lou's sister, Dorothy, came to visit. Of course we took her to all of our favorite nearby places, as well as to the Amalfi drive. We did Positano, Amalfi, and Ravello where we had lunch at a favorite restaurant. By the time she left Sorrento, we were worn out and were looking forward to our last hurrah for the season. It was to be an automobile trip to Calabria with Rakesh and Luisa. Lou and I had visited my relatives in the very south of Calabria, but we had never had the opportunity to visit La Sila, a mountainous plateau located farther north where it occupies part of the provinces of Cosenza, Crotone, and Catanzaro. This extensive region is rugged and is home to several peaks, the tallest of which is almost 2,000 meters. During the time of Mussolini, extensive hydroelectric developments were undertaken, and over the years, a substantial tourist industry developed. In the past, we had enjoyed car trips with Rakesh and Luisa in Arizona, so we thought that such a trip in La Sila might be enjoyable. We put the matter in Rakesh's hands, and he planned a memorable trip for us including hotel accommodations in Camigliatello, an almost alpine village in the heart of La Sila. Sant'Agnello was on the way for Rakesh and Luisa who left Naples and stopped to pick us up as we then proceeded south on the Autostrada del Sole toward our destination in Calabria. It was a long but beautiful drive, and since Rakesh was driving, I was able to take in the sensational countryside. When Lou and I had driven this way in the past, I was most always driving and could not dare to avert my eyes long enough to fully enjoy the scenery or the

Autostrada itself, a marvel in engineering. Our route took us to Cosenza and then upwards toward the highlands of La Sila Grande. We passed reservoirs, streams, and waterfalls much as we would see in our own western mountains. It was summertime when we passed mountains that are snow-capped in the winter. At last we reached Camigliatello, a tourist town, but not unpleasantly crowded. The city is surrounded by pine forests, and thus it was a real pleasure to sit on the balcony of our room to gaze at the trees and to try to identify some of the birds that flitted about. Unfortunately, I had not brought a bird guide with us, but I vowed that we would go there again for a longer stay and with binoculars and field guide.

In a typical Italian way, as the afternoon wore on we made our *passaggiata* looking at the numerous stores and the varied merchandise. It was amazing that this fundamentally tourist town was so uncrowded, but that was explained by the fact that mid-summer is low season in La Sila. This is the time of year when Italians go to the beach. High season in Camigliatello is in winter when skiing is the big attraction. It was no wonder then that there were so few patrons in our hotel and that its dining room was empty except for the four of us. The paucity of diners certainly had no negative effect on the quality of the food we were served. It was a copious meal with many courses. What I remember most clearly was a wonderful *penne con funghi* featuring porcini mushrooms. Calabria is famous for these mushrooms that abound in the moist forests that are everywhere. Since La Sila is regarded as one of Europe's great forest wonderlands, just imagine what a supply of mushrooms exists there.

Just outside of Camigliatello is La Sila's most famous woods, the Bosco di Fallistro. Here we followed a well maintained and well marked path past specimens of its famous trees, *giganti della Sila* (giants of La Sila), a kind of larch (*pini larici Calabresi*). These giant trees, which are over five hundred years old, six feet across and 130 feet tall, are truly spectacular and rival our own sequoias and redwoods. One of the pleasures of the walk was to see one of the denizens of this forest, gorgeous little black squirrels or, as the Italians say, *scoiattolo*. I am not sure whether I was more taken by these delightful creatures or by their Italian name.

Aside from the woodlands, there are beautiful open areas as well. One pleasant drive was around Lago Arvo that was surrounded by green fields. At this elevation, there were still red poppies in bloom. As we drove around the lake, and as lunchtime approached, we took advantage of a bar/restaurant that had outdoor tables. Their menu featured some Calabrian specialties that I was dying to sample. Calabria is famous for its prosciutto, salami, soppressata, and so on with all levels of spiciness. So we chose a platter of these great foods and downed them with bread and, of course, the common Italian beer, *Nastro Azzurro*. Now that was eating! Of course, we did visit some small towns and their churches, each of which has its own story. One of the best known and larger of those towns is San Giovanni in Fiore at the south end of La Sila. By continuing on the road past San Giovanni, it is possible to descend to the Ionian Sea and its beaches. After our three-day stay in Camigliatello, we made our way back to Naples to spend the night. The next morning, Lou and I began our voyage back to Tucson. Another magnificent sojourn on the Sorrentine peninsula was over.

Our frequent returns were not always based upon a scientific motive, but more because we were in Italy for some other reason. And, if we were going to be in Italy, why not visit a place we really like and see friends we cherish? Besides, we are comfortable with the culture. Some of the trips to Italy were in connection with Lou's work as a travel agent. She had become quite well known in the Tucson travel business as the expert on travel in France and Italy, and return visits to both countries were called for to broaden her own experience. It was during one of these trips that we found another haven that gave us much pleasure and that beckoned for a continuing series of visits.

During the spring of 1997, we felt the urge to see our friends in Naples and Sorrento. After all, it had been two years since our last visit! Lou agreed with the proviso that we plan our trip so that we could check on some destinations where she had sent clients, but had not seen herself. I was free to go when I wished after having formally retired in 1992. In late April, we flew to Milan where we rented a car and started our pilgrimage for the year. Our first destination was the *Cinque Terre* (five lands) on the Ligurian coast south of Genoa. This celebrated region is beautiful and, much like the Amalfi region, is situated on rocky slopes above the sea. The area is quite rugged and there is no good road access to its famous villages: Riomaggiore, Manarola, Vernazza, Corniglia, and Monterosso. The more hearty visitors do the towns by foot using a well-worn trail that ascends and descends, sometimes quite steeply. At the time of our visit, I was already having serious sciatic problems, but fortunately I was able to complete the walk from Riomaggiore to Corniglia before having to resort to the train.

We then headed south into Tuscany, first making a stop in the beautiful walled city of Lucca before arriving in the Chianti region south of Florence. After visits to the wine towns of Greve, Castellina in Chianti, and Rada, we descended into the Siena region. This is a gentler countryside of rolling hills of grasslands, fields of grain, and vineyards. Picturesque villages dominate many hillsides and others are the sites of farms with stone buildings. Allees of Italian cypress lead up to the main houses, and more of these beautiful trees often dot the grounds. Thinking about it reminds me of an earlier trip to Tuscany more than forty years ago when Lou remarked, "If Heaven doesn't look like Tuscany, I'm not going there!"

Next on the agenda was a well known inn and restaurant near the town of Sinalunga and then on to another town down the road, Montefollonico. Here we were booked to stay for two nights at La Chiusa, another small inn with a celebrated restaurant. When we arrived, Franco, the *portiere* (concierge), was waiting. Of course, we had a lovely room with a fine view across the valley to Montepulciano, a fascinating medieval hilltop city just six kilometers away. After a demonstration of how to make *pici*, a famous handmade pasta of Tuscany, we were ready for dinner which in every way justified the good name of La Chiusa and Dania, the chef and owner. We spent the next day absorbing the beauty of the region and visiting some of the historic towns of this part of Tuscany. As wonderful as was the previous night's dinner, we were not ready to again invest that amount of money for a second evening. When we asked Franco where else we could dine, his response was, "*in paese*" (in town). He directed us to a small hotel and restaurant, La Costa, that was owned and operated by a young man, Paolo Masini, then about forty years old. La Chiusa is actually outside the town walls, so in late morning we walked over to town and entered through an impressive medieval portal. We found La Costa and were greeted by Paola, Paolo's significant other, and later his wife. We were much impressed and made reservations for dinner that evening.

Montefollonico is a small medieval village whose origins go back to the thirteenth century and was founded as a Sienese fortress. As such, it was built on a hilltop and was surrounded by a wall that enclosed not only the town, but also the church, Pieve Di San Leonardo. Apparently there was a Cistercian monastery nearby, and it was the work of the monks that gave the village its name. It seems that they were engaged in fulling wool, and thus the name *Mons a Fullonico* was applied to the village. In briefly researching these facts, I realized that I had no idea what fulling was, so this called for further investigation. What I learned fits in beautifully with other tidbits of information that I have accrued since our first stay in Montefollonico. Fulling is the process of shrinking and thickening wool fabric with moisture, heat, and pressure. The process originated in Islamic societies in the Mideast about the ninth century, and is said to have been brought to Europe by the Knights Templar returning from the Holy Land. Their routes took them through Spain and Italy with a main route going through Bagno Vignoni, only about twenty kilometers from Montefollonico. Fulling seems to have flourished in the thirteenth century in Italy when Montefollonico was founded. This is consistent with the fact that, in this part of Tuscany, sheep raising is prevalent as a source for wool and cheese.

    The rocky promontory upon which the town sits has fantastic views. To the south, as I mentioned, is Montepulciano across a small valley, and to the east one looks across a large valley, the Val di Chiana. The length of the town extends along the steep edge of the promontory, and just below are green fields that extend into the distance. In historic times, the town and the fields made up a large farm or, as the Italians called it, a *fattoria*. Dwellings, barns, and storage areas were all built on the hill for protection. When Paolo bought the contiguous buildings that were to become his hotel, evidence of their having been part of the *fattoria* was abundant. Most impressive were the ancient ledgers in which the crops and yields were recorded. Paolo had undertaken a massive reconstruction of the buildings, doing much of the work himself. At the time of our arrival, he had completed five rooms for rental, a small restaurant, and a fabulous terrace that looks across the rolling hills of the Val di Chiana to the mountains on the other side. A thin silver line shimmering in the sunlight is Lago Trasimeno, just inside the border of Umbria, the adjacent province. The views at night are also wonderful as one sees the lights of hill towns across the valley. Among the brightest are the lights of Cortona, the town made famous in Frances Mayes' book *"Under the Tuscan Sun."*

    True to our sentiment, we did return to La Costa several times during the next thirteen years, and Lou sent many clients there as well. During this period, we became friends with the Massini family, including not only Paolo and Paola, but with Paolo's parents Rénato (René) and Emma, and with Paolo's younger brother, Terry.

    The spring of 1999 was especially eventful, for we rented a house in Montefollonico for four weeks, thanks to Paolo's help. When we asked if he knew of a place, he thought we might be able to rent the ancestral home of his electrician, Renzo. It was a large old stone house in a beautiful location in the center of town. Renzo and his wife, Claudia, lived outside the walls in a modern house while the old home was not being used. We could understand why, for the kitchen was a disaster. Nevertheless, there was lots of space, the bedrooms were comfortable, and there was a garden. Lou speculated about what she would do to restore the place for more comfortable living. We were both appalled in 2010 during our last stay at La Costa when Renzo asked us to come to see how he and Claudia had restored the home. They had seen the light and realized the charm of the ancestral home. It was not at all what we would have done, but everyone to his own tastes!

17

Green, Blue, Yellow, and White Eggs

During the more than twenty years that I maintained a breeding colony of Mexican leaf frogs, it became abundantly clear that it was important for me to have a hand in everything that went on. With my long experience, I was able to pick up on things that the various students, graduate students, and undergraduates alike, might have missed. In particular, I refer to variations of egg color that I observed. In nature and during the height of breeding season in our colony, the eggs are green. This is typical of the leaf frog family, and no doubt is a characteristic that evolved with egg laying among leaves. I knew that the green color of the skin was due to the mixing of blues and yellows, the former a product of light scatter from the iridophores and the latter from the yellow pigment of xanthophores. However, I had no knowledge of the situation with the eggs. A good opportunity to resolve this question arose in 1980 thanks to a winter visitor, Guido Marinetti.

When I was a boy growing up in Rochester, my best friend was Paul Francione. He had two older sisters who had each married brothers from the Marinetti family. Connie, the eldest, married Pat, and the younger sister, Antoinette, married Guido. I referred to Guido Marinetti in the preface when I gave an example of how the GI Bill afforded so many young men the opportunity to go to college. I really did not know Guido well until the fall of 1948 when Paul and I had each started our freshman year at the University of Rochester. At that time, Guido was also attending the University of Rochester as a chemistry major. Paul and I lived in the north part of town, while Guido and Antoinette lived even farther north. The University of Rochester campus was at the opposite end of town. Since Guido drove to class every day, Paul and I worked out an arrangement to ride to campus with him. I was very much in awe of Guido, this very serious man who was probably at least ten years older than we were. Our freshman year was pretty tough, and Paul decided to drop out. He ultimately moved to Detroit, and I pretty much lost touch with him. Guido did extraordinarily well and took his PhD in biochemistry at the medical school and then joined the faculty.

Some years later, I again had the opportunity to meet Guido when I had returned to Rochester to visit my family. At that time, I was working on our endocrinology text, and I needed to refer to some of Guido's publications. Somehow we got together and at the same time got caught up on family matters. Guido had become an authority on lipid chemistry and was held in high esteem in the medical school. I learned that Paul had done well and was working for Pan American Airlines, but was having problems with rheumatoid arthritis. It was not long afterwards that Paul took a disability retirement, and he and his family moved to Tucson. He was soon followed by Pat and Connie and their family. In short order, there was a small clan of Marinetti-Franciones in Tucson. Guido and Antoinette came to visit every winter and Lou and I joined them for dinner annually.

Guido and I, of course, talked science and soon began collaborating on the lipid composition of tadpoles. The work we did together resulted in the publication of three papers in 1981. Guido became intrigued by the problem of egg color. The next winter he set up shop in my lab, and we worked on the yolk pigments of *Pachymedusa*. The manuscript that resulted, "Yolk pigments of the Mexican leaf frog," was significant enough to be accepted for publication in *Science* in 1982. What we found was that, just as in the skin, the green color of the eggs was due to the combination of blue and yellow elements. The blue was due to the presence of bilivirdin, a hemoglobin derivative, and the yellow to a carotenoid pigment, lutein, an acidic xanthophyll. We published on the chemical characterization of these pigments the following year. For the next twenty years, we remained in contact with Guido and Antoinette even though they ended their trips to Tucson following the deaths of Pat, Connie, and Paul. Unfortunately, Guido passed away two years ago, well into his nineties.

With the establishment that green eggs result from the mixed presence of blue and yellow pigments, an explanation was provided for the observation I had made that often the first eggs deposited in the spring by females raised indoors were blue. Instead of having the opportunity to feed freely on an array of moths and other insects, such frogs were given a cricket diet. Initially, I purchased crickets commercially. These crickets had been raised on potatoes, and thus had had a diet deficient in carotenoids. This deficiency carried over to the frogs, and so, when the females were undergoing vitellogenesis, the laying down of yolk, the pigmentation of the egg lacked yellow carotenoids. This left only the bilivirdin component, and thus blue eggs were produced. As the number of frogs maintained indoors grew, the cost of buying crickets became exorbitant, and so I established a small cricket farm. In order to preclude the deficiency of carotenoids in these, my own crickets, I substituted carrots for potatoes and no longer obtained blue eggs unless I wanted them.

Among the frogs we raised indoors during the winter were froglets derived from tadpoles that had transformed in the greenhouses at the end of summer and early fall. Those that were fed potato-raised crickets suffered the same carotenoid deficiency as did the eggs, and thus had a decidedly blue cast. Their xanthophores lacked carotenoid pigments, but had enough pteridine pigments to render them a very pale yellow. There was not enough yellow, however, to produce green in association with the blue structural color stemming from underlying iridophores. By switching feed to carrot-raised crickets, these rapidly growing froglets became greener and greener.

One of the spawnings in my home greenhouse the next summer presented a real surprise. One morning, I was amazed to find a rather nice healthy young frog extruding yellow eggs. Obviously, she had not added bilivirdin to her eggs during vitellogenesis, and thus they were pigmented only by lutein. She had plenty of this pigment, since she had been feeding primarily on moths attracted to the UV light. I made a point to follow this female over the winter and ensured that she was fed only potato-raised crickets. When she again spawned early the next season, just as one would predict, she laid white eggs which lacked both bilivirdin and lutein. We raised some of these white embryos to maturity and, sure enough, about a quarter of these were also hemoglobin deficient. There was not enough time, nor enough hands, to follow up on these observations. In all likelihood, the hemoglobin-deficient phenotype was based upon a recessive gene, but I do not know. Moreover, I have no idea where in the hemoglobin/bilivirdin biosynthetic pathway the gene operated. I am sure that this whole system is too esoteric to command the interests of future scientists, and thus it falls into the realm of unfinished business.

There is another long list of unexplained or unresolved phenomena. One of these concerns the production of lutein. It is a given that many of the moths consumed by the frogs contain lutein, but they also contain other carotenoids including simple carotenes. It is well established that vertebrates do not synthesize carotenoids, and instead, they are obtained only through diet. However, vertebrates are obviously capable of metabolizing carotenoids in order to convert simple carotenes such as alpha- or beta-carotene to the more complicated xanthophylls such as lutein. For example, when we fed carrots to the crickets, they accumulated simple carotenes in their tissues. Somehow, the *Pachymedusa* that ingested these carotenes converted them to lutein for deposition in the eggs. Moreover, the conversion to lutein and its transport and deposition is hormonally controlled through the action of estrogens. An investigation of the molecular bases of these processes was beyond the scope of my laboratory and thus remains unelucidated. The same applies for bilivirdin, but we do know a little more. During vitellogenesis, estrogenic stimulation of the liver leads to the synthesis of the protein vitellogenin to which bilivirdin is bound, and the complex is carried in the blood to the ovary. The amount of bilivirdin being transported is significant enough to color the blood serum blue. In some simple experiments in which we injected either estrogen or testosterone into either male or female *Pachymedusa*, the serum became blue. At the same time, there was a drop in hematocrit, suggesting the involvement of the steroid hormones in hemoglobin metabolism. The entry of vitellogen into the egg (ovocyte) has been studied by others in the clawed toad, *Xenopus*, and we presume that this process, "cell drinking" or pinocytosis also holds for *Pachymedusa*.

I found this brief foray into the world of vitellogenesis and egg pigmentation fascinating, both for its own sake and because it fits in beautifully with a concept that I had earlier formulated in my own mind. The concept is essentially that an amphibian egg can be viewed as a kind of pigment cell. I published this concept in 1985 on the basis of the following: The eggs of most amphibian species are polar structures. The center of gravity of such eggs is close to the heavily yolk laden pole of the egg, the vegetal pole. It is generally unpigmented while the opposite half, the animal pole, is darkly pigmented by melanin granules just under the cell surface.

As is obvious from my accounts of our earlier years in Tucson, I spent a fair amount of time in the field observing where frogs live and what they do. In the moister canyons around town, one of the interesting denizens is the canyon tree frog, *H. arenicolor*. I often looked for it when hiking in Sabino Canyon in the Santa Catalina Mountains just north of town. At the lower elevations of the canyon, about 3500 feet, in late March or early April, the frogs bred noisily in quiet, shallow pools. Almost every year, I took my experimental embryology class to Sabino Canyon to collect eggs to use in various lab exercises. The animal poles of such eggs are a beautiful light brown in color. Later in the season, perhaps in May or June, when I often hiked in upper Sabino at about 9000 feet, I, of course, looked for *Hyla* eggs. The frogs also thrived at this elevation. Eggs were easily found and, unlike their lower counterparts, these had very dark animal poles. Since the frogs in this drainage, from top to bottom, are surely isogenic, the difference in animal pole pigmentation is undoubtedly due to environmental variations and, in particular, temperature. When I first made these observations, I concluded that the frogs at the higher, and thus colder, elevations were producing higher levels of MSH and thus accelerating animal pole melanogenesis. In later years, one of my PhD students, Phil Fernandez, actually measured circulating MSH levels in leopard frogs maintained at different temperatures. He found that MSH levels were indeed higher at low temperatures. The concept that the amphibian ovocyte is a kind of pigment cell is surely a venture into the esoteric. As such, it will remain untouched by future investigators.

18

Teaching

Thus far, I have presented an array of facets of my life in research, but given short shrift to the teaching side of my activities. It should be recalled that the original *raison d'être* for my hiring at the University of Arizona was to teach the required undergraduate course, vertebrate embryology. This I did, except for sabbaticals, every year from the spring of 1957 until 1994, two years after my formal retirement. In the early years, when the university enrollment was only 10,000 students or fewer, the course was offered only one semester per year. As enrollment grew and vertebrate embryology remained a requirement for zoology majors and for all premedical and predental undergraduates, I taught the course both semesters. I liked the teaching, and I was proud of the course I offered to the one hundred or more students every year. As I pointed out earlier, I had experienced superb training at the hands of Hans Holtfreter and Emil Witschi, and I extracted the best elements from each of their courses as I prepared my own.

My course was comprised of three fifty-minute lectures per week and two three-hour laboratory sections. In the early years, I gave every lecture; that meant a total of about thirty-six lectures per semester. In later years, I often brought in two or three guest lecturers, mostly so that the students could see a fresh face. I have always found it difficult to lecture from notes. Of course I prepared notes for my lectures, which I read before going to class, but I left them behind when I went to the classroom. In this way, I could give a smooth delivery, because everything I said was stored in my head, and it just flowed out as the fifty minutes went along. This method had the advantage of easily incorporating new knowledge into my classes each semester. I kept up with the literature and attended annual meetings, so I did not rely only on what was recorded in the course textbook. My method of extemporaneous delivery also allowed me to field questions from the students on the spot. I encouraged students to ask questions without fear of interrupting me. I realized, of course, that many students would not ask questions for fear of showing their own ignorance, but this is the way it is. Some students have confidence while others do not.

I have always found it amusing, and at the same time sad, that in order to teach in elementary school or in high school, a teacher must take a myriad of courses steeped in pedagogy. At the university level, on the other hand, it is taken on blind faith that a new PhD automatically qualifies a person to teach. I am not complaining, because somehow it seems to work.

The other face of the embryology course represented by the two laboratory sessions offered a much different experience and required different talents. Here the students were to study serial sections of embryos at different stages of development. Since they studied transverse, frontal, and sagittal sections of each embryonic stage, they gained a three-dimensional understanding of the developing embryo. For the first twenty or so years that I taught the course, each student had a compound microscope to view the various sections. Use of the microscope was another learning experience. The purchase and maintenance of the microscopes entailed a fair expense. This was added to by the purchase of glass slides of the various serial sections. In addition, slide breakage was also a problem. Some time in the late 1970s, the costs for microscopes and slides were such that a real strain was put on the departmental budget. Rather than risk the possibility of dropping the laboratory part of the course, I hit upon another idea. As part of my ongoing research program, I had purchased some excellent cameras and microscopes equipped with cameras. So I thought, why not photograph every microscopic section of an embryo and make them available as thirty-five millimeter transparencies. In this way, the problem of slide breakage could be precluded; moreover, if I chose the best slides to photograph and made multiple copies of each transparency, the students would have uniform sections to view, and they would be of the best sections.

In order to do all of this, I hired some excellent students to carry out the project using my lab equipment. We got rid of the microscopes and replaced them with Kodak carousels on desktop projectors. This proved to be an excellent way to teach the laboratory part of my course. The only real drawback was that the students were deprived of the opportunity to use compound microscopes. Since my short-term memory is no match for my long-term memory, I cannot recall for how long we used this laboratory teaching system, but I believe that it ended some time just before I retired.

One of the problems with teaching the laboratory portion of the course to so many students is that of recruiting teaching assistants (TAs). Some of my own PhD students served as TAs over the years, but often TAs were from other disciplines in the department. When they were accepted into the department graduate program, they were offered teaching assistanceships, and it sometimes ended up that they were to be associated with a course for which they had no particular expertise. It was just assumed that, because they were aspiring PhD students, they had enough smarts to pick up what they needed to know to teach a laboratory section. It was not the best system in the world and was somewhat unfair, but it worked out, and for my course I don't remember ever having a TA who failed his responsibilities. As the years went by and more and more courses dropped their laboratory offerings, the system evolved or degenerated, depending how one looks at it, and TAs received stipends without the need for a teaching experience. Graduate education had become more of a research endeavor and less of a teaching experience.

Perhaps I am a bit of a mossback, but let me expand upon my appreciation of the latter. About 1980, a new graduate student, Forest Morrisett, came to me in search of a PhD advisor. Forest made no bones about the fact that his aspirations were to obtain a PhD in order to undertake a career of teaching at the college level. Since he had obtained a master's degree from Arizona State University studying the comparative morphology of amphibian spermatozoa, he was sent to me as a possible mentor. By this time, the sentiment had developed that the PhD degree was one of research, and that students at this level were accepted by a particular professor in order to contribute to their respective research programs. As I talked with Forest, I realized that he was not at all interested in or qualified to work on any of the ongoing projects in my laboratory. Moreover, I could see no other of my colleagues who would be an appropriate mentor. In view of the fact that Forest's master's degree was in an aspect of developmental biology, I felt that it behooved me to accept him as my student. After all, he was a native Arizonan and wished to be educated at the University of Arizona, a state university. The question of a dissertation was another matter that could be dealt with in due time.

Forest took to his teaching role as a TA in my developmental biology course with great enthusiasm. He loved the students, and the students loved him. In fact, almost everyone loved Forest. Moreover, despite the fact that he was a little older than most of the other graduate students, he displayed a certain naïveté that made one want to reach out to him. He was a character, and he loved to talk about religion and philosophical subjects. It turned out that he was a natural teacher. Forest's progress toward a degree was slowed a bit by his personal life. He was a confirmed bachelor and lived together with his paternal grandmother whom he cared for until her death a year before he finished his degree. Forest's progress toward a degree was also slowed by his teaching activities for me; however, they led to an important creative contribution. He decided to write a laboratory manual to go with carousels of slides we employed in the laboratory sessions. It was an ambitious project, and I gave him free rein. I winced at his atrocious spelling, but this was corrected for the most part. It made me realize that intelligence and the ability to spell are not necessarily correlated. The finished lab manual was copyrighted in 1982, and I was proud to be co-author. The completed work, "Laboratory Manual for Developmental Biology, Cell 456," was spiral bound and sold in the university bookstore at a reasonable price.

Forest's PhD dissertation was another tome. He proposed to take advantage of the fact that there is no developmental staging series available for any of the phyllomedusine frogs and only a rudimentary one for the related tree frogs. Thus, in 1986 he completed his dissertation, "Normal development of Pachymedusa dacnicolor." I should have known that Forest would never write this work for formal publication. He did not, and I suppose he thought that I would; however, I never did for multiple reasons. Thus, this project remains another piece of unfinished business. As for Forest, that is another story and a fascinating one at that. Forest's desire was to teach, and thus he accepted a position at LaGrange College in LaGrange, Georgia, not far from Atlanta. This church-sponsored college seemed to be a natural fit for him; however, this was not the case, and after a few years he moved to a community college in Burlington, Iowa. Forest has found his niche and seems to be comfortably situated in Iowa. His life's story is one that I hope he will someday record.

19

The Developmental Biology of Pigmentation Patterns

Pigmentation patterns of vertebrate skin can be classified generally into groups: dorsal-ventral patterns in which the dorsal skin is darkly colored while that of the ventral side is light, and particular pigment patterns such as spots, stripes, or mottlings. Over the years, I have thought about these patterns from several vantage points, but especially from my perspective as a developmental biologist. Even as I was engaged in other aspects of pigment cell biology, as I have presented already, I began an examination of pattern development through the vehicle of teaching. In the spring semester for all my years of teaching, I taught a course in experimental embryology patterned after the course I took from Hans Holtfreter so many years before. As the years went by, I added new experiments that I thought would be of interest and of benefit to the few undergraduates and graduate students who took the course. I very much enjoyed teaching it, and I believe that the same held for the students who relished the intimate environment that derived from the presence of only a few classmates and me.

One of the experiments that we did was to make neural crest grafts using whatever species of embryos I had on hand at the time. I was an old hand at this based upon my early work with the salamander embryos. Sometime in the mid-1970s, I became quite curious about some work done by one of my friends at Tulane University, E. Peter Volpe. Pete was interested in some pigmentary variants of the common leopard frog, *R. pipiens*. The wild type had distinct, rather large dark spots on a tan or green background, while the *R. burnsi* variant had few spots, often as few as one or two, on a tan or bronze background. A third variety called *kandiohi* had spots on a mottled background that almost hid its spots. Pete and others had clarified the genetics of these patterns, and he had become convinced that these spotting genes operate at the level of the neural crest. As support for his conclusion, he reported that the results of transplantation of neural crest from one variant to the mid-ventrum of another resulted in a neatly discrete island of donor pattern on the white belly of the host. I had no argument with his conclusions, despite the fact that in the many neural crest transplantations that I had done on salamanders, I never observed such discrete islands of donor tissue; rather, the donor tissue was rather diffuse over the circumscribed area of the graft. In any case, I thought his experiments were rather simple and clear, so over a several-year period I had my students repeat them using wild type embryos. Indeed, my students got the same results as Pete, but as I observed this work closely, I was impressed by the fact that the discrete area of donor spots that developed on the belly differed from the dorsal spots in size and distribution. Moreover, when I looked at the resulting grafts carefully, I observed the presence of skin glands. Skin glands are normally present only on the dorsal surface and, moreover, they are not of neural crest origin. What the students had obtained was an island of dorsality. It dawned on me that what was transplanted in Volpe's experiment and in that of my students was neural fold that included not only neural crest cells, but skin cells as well. It appeared to me that the neat spots on the ventral surface represented the dorsal skin component of the neural fold, and that it dictated the pigmentary response of the neural crest cells that came along. Thus, the various pigment cell types differentiated and were cued by the skin to form spots. It would seem that this skin component of the graft already had the message that it was dorsal, but the message was incomplete and lacked the details of where the spots should be and of what size. Accordingly, I got a few of the more capable and interested students to do some follow-up experiments.

If my conclusions were correct that the prospective dorsal ectoderm already had the general message for dorsality, transplantation of a piece of this ectoderm excluding the neural fold, to the ventral side of a host embryo should result again in an island of dorsality with spots. In this case, the pigment cells of these spots would have derived from neural crest cells of the host that had migrated downward and, finding themselves in a dorsal skin environment favorable for their expression, went ahead and differentiated. To my delight, this is exactly what happened in those classroom experiments.

These results in turn begged the question of how and when it was that the large spots of the normal dorsal surface were determined. The "how it was" question was elusive, and I could see no means to solve it; however, the "when" question was more easily approached. We began to look for a solution by simply rotating pieces of dorsal skin on tadpoles of a manageable size. At this stage, the ultimate spot pattern is not evident and only becomes visible after metamorphosis. We rotated pieces of skin 180° and waited for the tadpoles to transform. At this time, it became evident that, at the time of skin rotation, the margins of the rotated skin patch often passed through prospective spots. Thus, after metamorphosis, when adult skin was formed, particular parts of individual spots were missing, but they could be found in all their precise details displaced by 180°. It appears then that, although spot or dorsal specificity of the skin is not yet determined at the neural fold stage, it has already been established in tadpoles well before they transform into froglets. I am grateful for the participation of several undergraduate students in this work including Mike Grossman, Larry Burstein, and Laura McCann. Forest Morrisett aided in the skin rotation experiments, but I cannot remember whether he was a student in the experimental embryology course or was acting as a TA. When these experiments were concluded, we were left with the unsolved problems of how it all comes about. Some insights became available from later work.

Although these experiments on the basis for the expression of the spot pattern in leopard frogs were initiated in my experimental embryology course, and much of the work was done by undergraduate students, it was original research that produced a definitive conclusion of some significance. As a result, in 1985, I submitted a short manuscript based on the work to the *Journal of Experimental Zoology*. The manuscript was accepted and published a few weeks thereafter.

One of my colleagues, Newell A. Younggren, a dear friend who had earlier been head of the Department of Biological Sciences, knew about our class project and asked if the work could be published in a different way. Newell was a teacher and, in fact, joined the department in the early 1960s to do just that. This was a time when there were considerable funds available for innovative teaching, and this was Newell's forte. Despite the fact that Newell was not involved in original research, he was chosen to head up the department because he was a capable academician and widely liked and respected. By the early 1980s when we were engaged in the spot pattern project, there had been a reorganization of biological sciences. Newell was no longer a head of department and was now primarily an advisor in the new Department of Ecology and Environmental Biology. My position had been transferred to the College of Medicine, and I was officially a professor in the Department of Anatomy. Nevertheless, I did not physically move to the College of Medicine campus, and I retained my research space in the same building, BioSciences West. My teaching obligations were unchanged. There was a remarkable camaraderie among some of us in the old department, and we continued to see one another despite the reorganization. It was thus that Newell knew about our leopard spot project.

Because of his interests in science education, Newell had an affinity for teaching publications such as *The American Biology Teacher* and the *Journal of the College Science Teacher*. He thought that our spot pattern experiments would be ideal for the *J. Coll. Sci. Teacher* if we described them in such a way that biology teachers could adapt them for class. I was happy to oblige, and thus we wrote a manuscript entitled "The expression of skin patterns in metamorphosing tadpoles." Newell, Forest Morrisett, and I were the authors, and it was also published in 1985.

I think back on Newell with considerable affection. Our group of departmental friends got together for coffee almost every day, and at the end of the week, we went off campus for "beer lunch." These were lunches of great frivolity as we all ganged up on Newell to tease him. He had nicknames for each of us. Fortunately, mine was simple: he called me Joseph, and I called him "Dad" because he was fourteen years older than I. Among the group, Sam Ferris, Pete Pickens, and I were the most constant. Others also came and went. One of the funniest examples of our ribbing concerned an action that Newell took at the behest of his secretary Ruth, a rather old battleaxe of a woman. At the time, miniskirts were in vogue. One of the young secretaries under Ruth's charge took to wearing a miniskirt that was rather becoming since she was tall, slim, and good looking. Ruth was very much offended by this attire, and she asked Newell to get it stopped. Newell told us about his interview with the young woman, and in the discussion it seems that she asked what the limits of the skirt length should be. Newell's response was that the buttock line should not be exposed. We exploded with laughter and named the buttock line "Younggren's crease." Whenever a short-skirted coed passed when we were with Newell, someone would ask, "Can you see Younggren's crease?"

Newell was a veteran of the Normandy invasion during WWII and was proud of his service. We also teased Newell about that because, some weeks after the invasion when he was on R and R in a hotel on the coast of Normandy, a contingent of German soldiers, who still occupied the Island of Jersey, carried out a raid on the mainland. Newell was rousted out of bed and taken prisoner in his pajamas. Thus, he spent the rest of the war on Jersey.

Newell was the first of us to retire, but we still continued our beer lunches at least once a month. As time went on, both Newell and Sam Ferris became widowers, but we still met for beer lunch. As others drifted away, the beer group was reduced to Sam, Newell, and me, but of course I no longer was able to be there in the summer when we were in Greer. Finally we began to meet every Tuesday, and either Sam or I picked up Newell since he no longer drove. In the fall of 2010, Newell could no longer come, and the day after Thanksgiving he passed away at the age of ninety-five. We lost a dear and much loved friend. Now, beer lunch on Tuesdays is reduced to Sam, now over ninety-two, and me, although Lou often joins us. She, too, was very close to Newell and holds a similar affection for Sam.

VCM and the basis of dorsal-ventral patterns

For a year or so after the sortie into leopard spot patterns, I gave the topic little more attention because, as pointed out earlier, I was very much involved with organizational affairs in the world of pigment cell biology. Hosting the XIIIth IPCC took up a lot of time, and this was compounded by the inauguration of the journal *Pigment Cell Research* (*PCR*). In 1987, I was able to get back into the pigment pattern business with the arrival of Toshihiko Fukuzawa. Toshi had just completed his PhD at Tohoku University under the direction of Hiroyuki Ide, who had been my post-doctoral student about ten years earlier. Thus, Toshi was a kind of academic grandson. His dissertation research concerned the basis of the dorso-ventral pigmentation pattern of *Xenopus*. Thus, his interests were in perfect accord with mine, and the activities in my lab were a good fit for the advancement of his career.

In Sendai, Toshi had shown that a factor present in the ventral skin of *Xenopus* had an inhibitory effect on melanophore expression and thus seemed responsible for its pale appearance. This putative factor was yet to be identified and was tentatively referred to as VCM (ventral conditioned medium), named after how it was obtained. Essentially, ventral skin was removed from young adults, chopped into small pieces, and left to stand in a physiological salt solution for a given period of time. Thus, the medium was conditioned by certain proteins that leached out. The conditioned medium was filtered through membrane filters and added to neural crest cultures for testing. A specific section of neural tube from *Xenopus* embryos of an early tailbud stage was removed, cleaned, and sterilized before being placed in a sitting drop of basic salt solution for culture. Daily counts of cells emanating from the tube were made, including melanized and unmelanized cells. The number of unmelanized cells in the standing drop cultures that had included VCM was remarkably less than in controls or in cultures that contained media conditioned with dorsal skin. It was concluded that the VCM contains a putative melanization inhibiting factor (MIF) that inhibits outgrowth and melanization of neural crest cells, and that this factor plays an important role in the establishment of dorso-ventral pigmentation patterns in amphibia.

Toshi was able to hit the ground running when he arrived in Tucson, because my *Xenopus* colony was ready to supply embryos and my lab was fully equipped for his needs. Immediately, we utilized the *Xenopus* neural crest system with various combinations of VCM and MSH. The VCM was derived from leopard frog skins, and we found that it worked as well as that from *Xenopus* skins. Soon, with the help of Kristie Kreutzfeld, we found that VCM inhibited melanization of cultured mouse melanoma cells. Kristie was my lab technician who had previous experience with cell culture. She had been my right hand for several years after having worked at the University of Arizona Cancer Center. Kristie was the mother superior of the lab at that time, as well as being my assistant as I began work as editor of *PCR*.

In addition to working with the pale ventral skin of various frogs, we began to use dorsal skin conditioned media, especially spots versus interspots. Most of this work was done by an undergraduate student, Francesco Mangano. In this work, we found that dorsal skin conditioned medium, especially that from the dark spots, had a stimulatory effect in the neural crest explant system. Thus, it became obvious that dark skin contains substances that stimulated melanization while pale skin substances inhibited the process. With the help of a visiting scientist from Spain, Adelina Zuasti, and my graduate student, Warren Johnson, we extended this work to catfish. Dark dorsal skin of the catfish also possesses stimulators of melanization while in the pale ventral skin, inhibitors dominate. In his PhD work, Warren worked with hooded rats and showed that black skin contained melanization stimulating activity whereas white skin condition media inhibited melanization.

It was now time to try to purify MIF and to elucidate its identity. To this end, I enlisted the help of my friend, John Law, who was head of the Department of Biochemistry. John suggested that I hire one of their new PhDs to work on the project. He was Preminda Samaraweera, a very nice young man from Sri Lanka. He and Toshi did most of the work. VCM from one of the leopard frog species was purified, and fractions were tested with the neural crest explant system. Monoclonal antibodies against purified MIF were prepared for attempts at final purification by immunoaffinity chromatography. The monoclonal antibody was also used for immunohistochemistry. Preminda was not able to complete the biochemical elucidation of MIF for various technical reasons, so this work was abandoned. My assessment was that direct attempts at purification were too difficult and that a better approach would have been to employ the techniques of molecular genetics. This was well beyond my sophistication, and given the fact that retirement was in the offing, I abandoned this part of the project and left it to Toshi to complete when he returned to Japan. In the meantime, we went ahead with the use of the monoclonal antibody to localize MIF in the amphibian skin through the means of immunohistochemistry.

This next aspect of the problem was aided by a discovery that I made on a side project that I was carrying out with Kristie. For reasons that I'll not explain here, I was interested in the possibility of obtaining primary cell cultures of iridophores, those flashy reflecting pigment cells that I had studied for years. Given Kristie's expertise with cell culture of mammalian pigment cells, I asked her to try to culture iridophores from leopard frog tadpoles. It didn't take long before she obtained primary cell cultures of pure iridophores that were indeed beautiful. The cells did very well and firmly attached themselves to the bottoms of their culture dishes. We began to treat them with various agents. When MSH was added to the culture medium, the iridophores responded by rounding up just as they do in the skin of a living tadpole. Of greater interest to me at the time was that, in cultures containing VCM, the iridophores spread out fully, just as they do *in situ* on an hypophysectomized tadpole. Thus, it would seem that

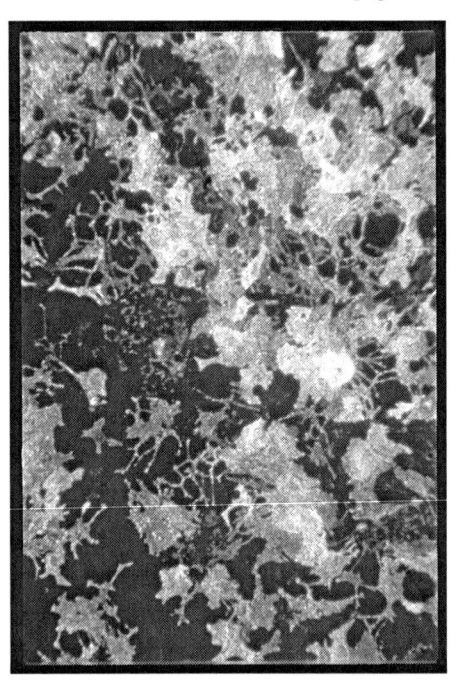

Primary cell culture of iridophores exposed to VCM.

the action of VCM, through its constituent MIF, serves to make the ventral surface lighter by both inhibiting melanophore expression and enhancing that of iridophores. If this reasoning were correct, the use of the monoclonal antibody against MIF in immunohistochemical experiments should result in strong staining with the antibody in the ventral integument where iridophores abound. This was indeed the case, providing further evidence for the important role played by MIF in establishing and maintaining the pale pigmentation of the ventral surface.

  This is where we were in the early 1990s. We were not able to provide a chemical elucidation of MIF, but the evidence for its existence was indisputable. I hated to speak of the "putative MIF," but that is where we were.

It was time for Toshi to return home after having spent almost four years with me.

Toshi was from Nagano in the mountainous region of central Japan, where his parents owned agricultural land. As I later saw, it was almost an urban farm. Just before he left to come to Tucson, his father passed away, leaving his mother to tend to the land. I am sure that Toshi was concerned for her. Moreover, he needed to advance his own career. I do not remember the exact chronology, but about this time a faculty position opened at the Yokohama/Hiyoshi campus of Keio University, and Toshi became a candidate. This was the department of Jiro Matsumoto and Mat Obika who were my friends. I was also friendly with other members of the department. Whether I had any influence, I do not know, but in any case Toshi was awarded the position and thus was the third member of their faculty to have been "contaminated" by a stay in my laboratory. I have since liked to tease my Keio friends by referring to them as the "Tucson mafia" of Keio University. With Toshi firmly ensconced at Keio, I was happy to give him the wherewithal associated with our MIF research, including the monoclonal antibodies. Thus, I could clear my conscience about the possibility of this project becoming another piece of unfinished business. The project was now in Toshi's hands.

I cannot leave this topic of putative MIFs and MSFs without giving them their due credit of importance. They are certainly involved in the expression of the most widely distributed of all pigmenting patterns, that of dorsal versus ventral. Even humans express such a pattern, although it is not widely appreciated. Many of the unexplained human hypo- and hypermelanoses may be caused by an imbalance in putative MIFs or MSFs. I have raised this possibility many times at meetings of the pigment cell community, but it has fallen on deaf ears. Dermatologists generally look at the gap between lower vertebrates and man as too great to be possibly bridged. As one last jousting with a windmill, I undertook a small collaborative study with my friends in the biochemistry department at the University of Murcia (Spain). I provided Paco Solano and his colleagues with some of our partially purified MIF. They found that MIF exerted an inhibitory effect on the activity levels of two important enzymes in melanogenesis, tyrosinase and dopachrome tautomerase, when tested on two mouse melanoma lines. Moreover, MIF blocks the stimulatory effects of MSH on these enzymes, probably at the level of the MSH receptor. MIF seems to have the same mechanism of action as does agouti protein, a factor of influence in regulating coat color variants of mammals. Perhaps someday, someone will grasp the relevance of these results. In the meantime, it remains as more unfinished business.

20

The Final Years of Research

The end of the 1980s and the beginning of the 1990s was a period of intense activity in my laboratory. I have described some of the research when I covered our exploits with the Mexican leaf frog and our occupation with pattern formation. As time went on, I was less and less involved in the everyday experiments because I had good help, between Kristie and my students, in getting jobs done. As I pointed out, Kristie's success with primary cell cultures of iridophores was an important help in the MIF experiments. The availability of these cells was helpful in answering other questions about iridophores that had arisen over the years. One of these concerned the exact composition of iridophores. We were pretty certain that their crystalline components included guanine and probably other purines. It was also possible that pteridines in the crystalline form were present in these cells. The availability of pure iridophores in quantity sufficient to ascertain their various constituents provided us the means to answer the question. Again, my friend Al Cohen provided his facilities and expertise as we analyzed the iridophore preparations for purine content. We were able to show that guanine was the exclusive purine in these cells. Phil Fernandez, one of my graduate students, did a chromatographic analysis of the iridophore extracts and demonstrated the presence of two pteridines, isoxanthopterin and amino-4-hydroxypteridine. Obviously, guanine and the pteridines were present in crystalline form, but whether they are present as individual crystals or as mixed crystals remains unanswered.

During her tenure in my lab, Kristie participated in other projects that I'll not discuss here at length because they were unrelated to my main research objectives. One of these was the development of a refined radio-immunoassay for MSH that was also applicable to the detection of the super potent analog (NLeu^4, D-Phe^7)—α-MSH. We did this because I hated the frog skin bioassay that Mac Hadley utilized which involved decapitating frogs to obtain their skin. Kristie's work on this project made me even more aware of her talents in carrying out original research. She certainly would have made a success should she have chosen to pursue a PhD. I encouraged her to do so, especially since I knew she was envious of my PhD students and those of Mac Hadley. However, she chose instead to become a chiropractor. She left the lab with my blessing, but she kept in touch as she completed her chiropractic degree and began a successful practice in Cottonwood, Arizona, that I believe still continues.

One of my PhD students during the late 1980s was Wade Sherbrooke. He was unique in that, while I was his major professor and dissertation director, he was essentially independent. Wade was older than my other students and was already an experienced herpetologist and ecologist. After an earlier stint as a graduate student at the University of Arizona, he had left for the Peace Corps. He returned to the University and was employed by the Arid Lands Program. I got to know him during this period when he asked to use my microphotography facilities to photograph pollen grains. During this time, he realized that he needed a PhD degree to better insure job security. Rather than return to his old herpetology mentor, Chuck Lowe, he discussed a return to graduate school with Newell Younggren. Wade had already begun independent research on horned lizards and wished to pursue this work for a dissertation. Newell immediately suggested that he see me about taking him on as a student. I was happy to do so, for I knew that Wade was exceedingly capable; moreover, I knew that I would be happy to enjoy vicariously his research on horned lizards. Wade set himself up in the lab, became a TA in my embryology course, and generally contributed to the research ambiance of the lab. Of course, I followed his research as closely as I could, but he was really independent. A string of excellent papers flowed from his dissertation research as Wade became the world authority on horned lizards. After he completed his degree, he was offered the directorship of the American Museum of Natural History field station in Portal, Arizona. During his tenure at the Southwestern Research Station, he obtained significant research support for the station and added to the facilities and to the various programs and services it provided.

For many years, I had been interested in the role of MSH in the expression of pigmentation patterns. I knew that this hormone could not determine exact patterns for that was the realm of the interaction of neural crest cells with the tissue environment they found themselves in. However, by stimulating or inhibiting the various pigment cells or propigment cells, patterns could be markedly affected. One of the best examples of this effect was revealed in the MS thesis of Phil Fernandez which he completed at Arizona State University. Phil was struck by the profound differences in appearance of adult tiger salamanders (*A. tigrinum nebulosum*) depending upon which forest pond they came from. It seems that, on background that favored MSH release from the pituitary, melanophores were stimulated, and a dark pattern was expressed. In marked contrast, turbid ponds provided a light colored background unfavorable for MSH release, and thus, salamanders found there were light colored because iridophores flourished, and the dark pattern was obscured. At some point, I met Phil and showed interest in his work. Subsequently, he applied to our program and came to the University of Arizona to pursue his PhD under my direction.

Given Phil's background and interests, I acquainted him with a pigmentation problem that I knew would intrigue him. It concerned the probable role of MSH in regulating the morphological expression of pigmentation pattern in the leopard frog, *R. chiricahuensis*. Unlike other sibling species of leopard frog which have a markedly pale ventral surface, *R. chiricahuensis* has the capacity to express a heavily melanized ventral surface. This effect is especially manifested in the winter when the frogs are torpid near the bottom of their pools. Not only does the ventral surface become dark, but the same is true for the dorsal surface that becomes so dark that the spot pattern is barely visible. Phil took to the problem for his dissertation work. He showed that the darkening was due to profound elevations of circulating levels of MSH induced both by low temperatures and by maintenance on a dark background. In fact, by the use of the radioimmunoassay for MSH that Kristie had developed a year or so earlier, he found that the levels were eight times higher in *R. chiricahuensis* than in *R. pipiens*, another leopard frog, under the same circumstances. Two solid original research publications derived from this work.

Although Phil must have sensed a certain amount of pressure from some of my faculty colleagues in the Department of Anatomy to take a post-doctoral fellowship toward a career in research, Phil chose another route. He had a strong interest in teaching and, moreover, he wanted to live in Phoenix, his home town. Thus, when he was offered a faculty position in biology at Grand Canyon University, there was no hesitation on his part to accept. In short order, he launched a very successful career as a teacher, administrator, and part-time researcher in this small liberal arts college. We were pleased for Phil and his wife Cindy. As their family grew, we stayed close friends, and indeed Lou and I love this Fernandez family. Phil now has a new job after commercialization changed Grand Canyon University. He is chairman of the Biology Department of Glendale Community College and is highly respected by his faculty and the community. Phil and Cindy's children, Emma and Philip, are grown and are in college. We are happy for the Fernandez family.

Another student from these final years for whom I have great affection is Francesco Mangano. Francesco was not a former graduate student, but an undergraduate who contributed significantly to our research on pigmentary pattern formation. I remember well one day in early 1990 when a nice-looking young man came into the laboratory and asked if he could speak with me. He was quite polite and neatly dressed in a shirt and tie. He was a premedical student who was looking for the opportunity to participate in research in order to enhance his credentials for admission to medical school. After a short conversation, I was impressed enough to allow him the opportunity to work with us. In short order, I realized that I had made a good choice in this young man, Francesco Mangano, for he was bright, serious, reliable, hard working, and very pleasant. I learned that he was born in Italy and that his family had immigrated to this country when he was a teenager. His parents were from the south, his father from Sicily and his mother from Puglia. They, like so many young people, had moved to northern Italy to find work. They met and married in Torino and settled there. His father became a machinist, an occupation that brought him and his family to the United States where they became successful citizens. Francesco obviously came from a good home and had acquired a good work ethic. In the lab at this time, Toshi and Warren Johnson were hard at work using the *Xenopus* neural crest explant system, and they taught it to Francesco. It was thus that he undertook the work on MIF and MSF in leopard frogs that I mentioned earlier.

During his stay in the lab, Francesco applied for admission to the University of Arizona medical school. He was not accepted on at least two attempts before giving up on the University of Arizona. Not only was he disappointed, but so was I. In fact, I was furious and frustrated because this was another example, among many, of what I consider to be a flawed admission system. Here again was a bright, capable, and compassionate young person who was denied admission while many of his peers with less redeeming attributes were accepted with ease. In order to accomplish his goal, Francesco accepted admission to an osteopathic medical school and went on to a residency in pediatric neurosurgery. He was a successful researcher in this area and has become an established pediatric neurosurgeon in Cincinnati. During his residency in New York, he met Danielle, a stunning young woman whom we met in Tucson just before they were married. They have produced four beautiful children with lovely names to match their good looks. We are close to Francesco and Danielle and are so happy that they keep in touch with us.

21

Japan and Friends

In the course of this narrative, the various scientific projects of my Japanese colleagues, post-doctorals, research associates, and visitors have been touched upon, but that is only part of the story. Many of those named have become close friends over the years, and my personal interactions with them, both here and in Japan, are memorable. It all started in 1961 when my first visit to Japan took me beyond the endocrine conference in Oiso, to Tokyo and Yokohama. The latter was to the Yokohama/Hiyoshi campus of Keio University. Here, I first met Tadao Hama and his young associates, Mat Obika and Jiro Matsumoto. Little did I know that Mat and Jiro would become life-long friends. Hama, it turns out, was the patriarch of a whole lineage of young Japanese investigators who came to my lab in Tucson. Mat and Jiro, of the first generation, were not really students of Hama, but he was their boss at Keio, and their research work was concerned with Hama's projects on pteridine pigments of amphibians and fishes. Subsequently, Hama moved to Nagoya University while Mat and Jiro remained on the faculty at Keio. In Nagoya, a second generation of Hama students arose. Two of these young scientists, Hiroyuki Ide and Masumi Yasutomi, each spent about a year in Tucson. The last of this "Hama school" of investigators was Toshihiko Fukuzawa, my last Japanese associate, who was actually a PhD disciple of Hiro Ide who had joined the faculty of Tohoku University in Sendai after leaving Nagoya. Over the years, Lou and I have had the honor to be guests in the homes of each of these young people.

Not all of my Japanese personal associations were related to descen-

Japan Pigment Cell Club at Kegon Falls in Nikko in 1969.

dants of the Hama school, for many others came from the worlds of pigment cell biology, comparative endocrinology, and developmental biology. Many interactions were related to scientific meetings or conferences that required travel to Japan. As I look back, I realize that, starting in 1961, I have made twelve visits to Japan. These visits provided the wonderful opportunity to see much of Japan, from Kyushu to Hokkaido. Of course, the majority of time was spent on the main island, Honshu.

After my 1961 visit, there was not an opportunity to return to Japan until 1967, and that trip was only in passing on our return from India as I mentioned earlier. In 1969, I was invited to participate in a US-Japan seminar held in Tokyo. It was held in conjunction with the annual Japan Pigment Cell Club meeting that followed in Nikko. There was a day between the two events, I believe it was a Saturday, and this allowed me to see Mat and to meet his wife Yoshiko and their little boy. Mat and Yoshiko were living in a separate part of the house of his parents in Kamakura, home of the Great Buddha (Daibutsu). I visited them there, and this was my first visit to a Japanese home. I looked forward to meeting the family for several reasons; one was that Mat's father must have been a very important person at one time. I did not learn this from Mat, but someone had told me that during the war Mat's father had been an admiral in charge of the naval installations in Tokyo Bay. Be that as it may, Mat's father looked like any Japanese gentleman of his age. He seemed very nice, and Mat's mother was sweet. They were just ordinary people. I had not given the matter much thought, but at the time of that visit, the war with Japan had ended just twenty-five years earlier. Now, we were all friends who enjoyed and respected one another and their respective cultures. In fact, on that day, Mat and I shared a pleasant mingling of the two cultures by doing what is now referred to as "a guy thing."

After the war, the Japanese adopted baseball and, in just a few years, it became a passion, attracting fans from all walks of Japanese life. By chance, the Japan Series was in progress, and Mat was all for sitting in front of the TV to watch. It seems that Mat had won a case of beer in a bet with one of his friends over the outcome of an earlier playoff. Thus, I soon became educated in Japanese major league baseball over a few bottles of Kirin, my favorite Japanese beer at that time. Mat's favorite team, the Tokyo Yomiuri Giants, was playing, but I don't remember the opponent. The Tokyo Giants were the equivalent of our New York Yankees and featured a player who later became renown, even in the USA. He was Sadarahu Oh, who broke the long-standing home run record of sixty. Interestingly, I remember more about that baseball session than I do about the meeting held in Nikko on subsequent days.

My second opportunity to visit a Japanese home came in 1977 when I was in Tokyo for the International Developmental Biology Conference mentioned earlier. I was taken by Jiro to their home in Yokohama. Jiro's father was still alive at the time, but he was living in a separate part of the house while Jiro and Akiko occupied the majority of the dwelling. In subsequent years, we got to know the Matsumoto's dwelling rather well as houseguests during several visits. We also got to know the neighborhood pretty well as we walked with Jiro and Akiko. What stands out in this regard is a beautiful Zen Buddhist temple that we have visited on many occasions.

Aside from the association with the Hama school, an unrelated Japanese connection started unexpectedly in Tucson in 1968. When Chris Richards, who was with me in 1965, returned to her home in Michigan, she took on the position of directing a new amphibian breeding facility at the University of Michigan. I had recommended her to an acquaintance in the faculty there,

Dr. George Nace. George had received support from the National Science Foundation to establish a facility to maintain genetic strains of the leopard frog, *R. pipiens*. The facility was to operate much like the famous national mouse laboratory at Bar Harbor, Maine, that made available specific strains of laboratory mice for use by American and foreign investigators. George's frog facility had a counterpart in Hiroshima, Japan, that had been started by Professor Toshijiro Kawamura, president of Hiroshima University and an important person in Japanese science. In the fall of 1968, Professor Kawamura was visiting George in Ann Arbor. He indicated a strong desire to visit the Grand Canyon and other important sites in Arizona. George called to ask me to help. I was not too keen about doing much since the fall semester was well under way, and I knew that Professor Kawamura's use of our language was limited. The latter was not a problem for George, who was a fluent Japanese speaker. However, since Jiro Matsumoto was then in my lab, he and I could take Professor Kawamura to northern Arizona. Off we went to the Grand Canyon, and while it went well and Professor Kawamura was quite pleasant, I limited our tour to only the Grand Canyon. After his return to Japan, we stayed in touch until his death a few years ago. We had in common an interest in amphibian husbandry. Moreover, pigmentation was an important parameter in his laboratory's work on various mutants.

During the summer of 1979, when I was to be in Japan for an extended period as a Research Fellow of the Japan Society for the Promotion of Science (JSPS), Professor Kawamura invited me to spend a few days in Hiroshima to visit the facility and to give a lecture on pigmentation. By chance, George Nace and his wife were also in Japan as he, too, was a Research Fellow of JSPS, and the three of us were guests in Hiroshima at the same time. It was a fascinating visit for many reasons, made even more interesting because of George's facility with the language. I asked George where he learned Japanese, and he related a remarkable story. George was from Pennsylvania, born to Quaker parents. When he was a boy, his parents served as missionaries in Akita, a small city in the Tohoku District in the northeast of Japan. His playmates were Japanese, and thus he learned their language. When his family returned home, he gradually lost his ability in Japanese as he continued his American education. During WWII, George was drafted into the Army, and somehow it was revealed that he had spoken Japanese as a boy. The Army sent him to the Monterey language institute, and very quickly the Japanese language, hidden in the recesses of his brain, emerged. He soon became proficient again, so much so that when the USA occupied Japan at the end of the war, he accompanied General Douglas MacArthur to serve as a translator. One of his jobs was to interrogate Japanese prisoners. In doing so, one day George realized that he knew the man he was interrogating. He asked, "When you were a boy in Akita, did you play with an American boy?" The response was, "Yes, Georgie san," to which George replied, "I am Georgie san!" Despite George's excellence in Japanese, he did not seem willing to lecture in the Japanese language. I never knew why, but perhaps it was due to something I learned from Japanese colleagues who were both impressed and amused by George's facility in Japanese. It seems that when George's Japanese reemerged, what came with it was the thick accent of the Tohoku District.

Another highlight of the Hiroshima visit took place one evening after Professor Kawamura had taken us to dinner. One of the other guests was another Kawamura, a much younger man unrelated to Professor Kawamura. Since he was young, he still had black hair, so George and his wife referred to him as "black Kawamura" and to the much older, white-haired Professor Kawamura as "white Kawamura." Black Kawamura had been invited to the dinner since he had known George and his wife in Ann Arbor when he had been there as a student in physical education. After the dinner, black Kawamura invited us to attend a game of the Japanese Professional Baseball League. Black Kamamura was the official trainer for the Hiroshima Carp who were that evening playing, if I recall correctly, the Tokyo Yakult Swallows. Kawamura took us to a special dugout, adjacent to the players' dugout. Here we met the wife of one of the American players on the Carp team. In those days, each team in the league was entitled to sign on two American players. Unfortunately I cannot remember the name of the player, but it was one I recognized at the time. We had a pleasant visit with the wife during the course of a few innings before we left. It was a memorable end to a remarkable few days in Hiroshima.

Research Fellow, Japan Society for the Promotion of Science

In 1972, when I was in Tokyo for the International Developmental Biology Conference, I again met Professor Kawamura. He was attending the conference with his long-time associate, Dr. Midori Nishioka. It was she who ran the every-day operation of the Hiroshima amphibian facility. Either in conversation there or in later correspondence, I learned from Professor Kawamura about the existence of the JSPS, and about the possibility that it might provide the means for me to come back to Japan in order to visit the Hiroshima facility. After the meeting was over, he sent Dr. Nishioka to the JSPS offices to inquire about the possibility of inviting me to Hiroshima. They had already asked JSPS to do so for George Nace, and they were informed that they could not have two invitees. As an alternative, Professor Kawamura asked Professor Goro Eguchi of Nagoya University to serve as my host. I had met Goro in Tokyo and already knew his work on development very well. He did the footwork necessary for an application to JSPS and worked out a strategy that would lead to a favorable outcome. Eguchi had been a Professor at Nagoya for only two years, and he felt the application would be strongest if the official host were Professor Makoto Seiji, head of dermatology at Tohoku University in Sendai. Professor Seiji was extremely well known and moreover was a pigment cell researcher. Thus it was decided that my forty-five-day visit would be hosted by Professor Seiji, but that it would be tripartite with stays in Nagoya (hosted by Eguchi), Sendai, and Tokyo (hosted by Jiro Matsumoto). The application was successful, and I was awarded a JSPS Fellowship to start about June 1, 1979.

Actually, my stay in Japan began on June 4 when I flew from the US to Sendai via Tokyo's Haneda Airport. There, at Sendai, to meet me were Seiji, Takeuchi, Tomita, and Oikawa. After some organizing discussions, I flew to Hiroshima to start the Nagoya segment. I have already related the story of my experiences in Hiroshima that preceded my visit in Nagoya where I arrived by train for a whirlwind segment of the itinerary. My first evening in Nagoya gave me an interesting glimpse of Japanese life. After going out to dinner with Goro Eguchi and his wife, we went to a karaoke club. They were regulars at the club, and they kept their own bottle of Suntory whiskey in storage there for subsequent visits. The formal visit was with Eguchi at his Institute of Molecular Biology of Nagoya University. I also made a side trip to Kobe to give a lecture for my friend Yutaka Mishima and his group. About then, Lou arrived in Osaka from Tucson, and we proceeded to Kyoto for a lecture and for a couple of nights before returning to Nagoya. After Lou arrived, the Eguchis had us to dinner at their home and were most gracious.

We were happy to see two of my former Tucson guest researchers, Hiro Ide and Masumi Yasutomi, who were installed in Nagoya along with Tadao Hama. Hiro and Masumi took us on some sightseeing in Nagoya, and on our last day Lou and I took the train to Kyoto and Nara for our own sightseeing. The following morning, we awaited Hiro who was to take us to the airport for our flight to Sendai and the next segment of the stay. We had descended to the lobby of the hotel with our baggage and put it aside while we waited. Suddenly, Lou asked, "Who is that man by our baggage?" To my amazement, there was Tadao Hama whom neither one of us had seen since his visit to Tucson in 1961. We walked over to greet him and had a brief and halting conversation. He was such a shy person. He asked Lou if she liked Olympus cameras, and she replied that she knew little about them. He said that Olympus had just come out with a new and novel model, but that it would not be available for another week or more. He asked where we were staying and when we were leaving Japan. He then announced that he wanted to give the new model of Olympus to Lou because of our kindness to all his Japanese protégés who had spent periods in Tucson. Hiro arrived, and we bade farewell to Nagoya.

The Sendai segment

Although Makoto Seiji was our official host, Takuji (Taki) Takeuchi was our day-by-day host, and it was he who met us at the airport. I knew Taki pretty well; in fact, I had already visited him in Sendai in 1977 after the Tokyo Developmental Biology Conference. Moreover, Taki and his wife, Cheiko, had visited us in Tucson with their young son in 1978. That year, he had been a guest in Walt Quevedo's lab at Brown University. During their return trip to Japan, they stopped in Tucson. Little did we know that they would return the hospitality just a year later.

I hit the ground running when we arrived in Sendai. One of the first orders of business was to meet with Seiji and his colleagues who were serving as an organizing committee for the XIth IPCC to be held in Sendai the following year. I am not sure that I helped very much, for they had already come up with a solid program. These discussions were followed by my seminar on developmental aspects of pigment cells. After a visit in the lab with Taki Takeuchi, we had the pleasure of some wonderful sightseeing. Among the most impressive of the sites we visited was Matsushima, a "pine-clad island" that is considered to be one of Japan's most impressive scenic places. Matsushima is the name given to the whole area of Matsushima Bay, some fourteen kilometers long and twelve kilometers wide. It forms the innermost part of Sendai Bay, and it contains numerous small islands that are each covered with pine trees. I had been given a brief glimpse of Matsushima in 1977 when Taki took me on a quick visit of the Bay. I had a different view this time on a more extensive tour that we took with Seiji Sato, one of Taki's associates. We took a pleasure boat, decorated in the form of a dragon, that provided a changing view of the various islands. One year ago, Matsushima was in the path of destruction wrought by the infamous tsunami that took so many lives and devastated so much property. Lou and I were saddened when we heard of the tragedy, but took some solace when we learned that a curvature in the shape of the Matsushima Bay had protected the little islands from damage.

The next day, we toured by car with Taki and Cheiko. A memorable event of this trip was a visit to the village of Togatta, one of the places where Kokeshi dolls are created by noted artisans. These wooden dolls, turned on lathes, are fashioned from hard woods of the area, including maple. They are about eight to ten inches tall and are generally cylindrical, but are topped by an attached roundish piece that forms the head. The noted artists who produce them outline the facial features with fine lines of black paint. Each face is distinctive. The cylindrical bodies are decorated with red painted designs. Each doll is unique and the best dolls are signed by the artist. We watched a master Kokeshi doll maker and his son at work at their studio in Togatta. While the son was explaining the process to Taki and Cheiko, the master quickly turned out a lovely little top as a gift for Lou. Of course, we all bought signed Kokeshi dolls. It seems that this was a regular thing for Taki, who had a large collection at home. We had the opportunity to examine the collection when we had dinner at the Takeuchis. We were so taken by these beautiful little dolls that we have our own small collection comprised of either dolls that we purchased or were gifted us by Taki.

Hokkaido

The next day we flew to Sapporo on the north island, Hokkaido. Our host there was to be my old friend, Ryozo Fujii, whom I had known since 1961. I believe that we were there during a bad time in Ryozo's life, and we soon realized that there were problems that were translated into his ability to be an effective host. He had arranged for us to stay in a dormitory that was pretty dismal, compounded by the fact that toilets were down the hall and were of the "slit trench" variety. We thanked him, and proceeded to a more suitable downtown hotel. We then called my friend Kowichi Jimbow at the medical school. Kowichi had been indicated as a secondary host. I knew Kowichi quite well since he had spent about five years in Tom Fitzpatrick's lab at Harvard. From this point on, Kowichi felt free to take on many of the hosting functions for the rest of our visit. Of course, we were pretty self sufficient as long as signs were written in English.

We were alone our first night in Sapporo, but, as luck would have it, we encountered a young clerk at the front desk of the hotel who was pleasant, friendly, and quite fluent in English. We discussed with him where we would have dinner, and he suggested that the Sapporo brewery would be an interesting place to go and that they had an unusual dinner opportunity. Their featured meal was Genghis Khan, what we now know as Mongolian barbeque, and that they offered "so much ram," as the young man said. Of course, "ram" meant lamb. It sounded intriguing to us, so when the young man offered to take us there, we invited him to join us as our guest. I had an ulterior motive in asking him along because we were anticipating a problem over the coming weekend when we were to be alone at the end of our stay in Sapporo. We thought that we might take a train to a beautiful lake in another part of Hokkaido; however, unlike Honshu and other parts of Japan where direction signs were indicated both in English and Japanese, such apparently was not the case in Hokkaido. My thought was that we might be able to hire the young man as a guide. The evening was great, and the Genghis Khan was good, fun, and interesting. The meal was helped by the fact that it came with free beer, and both Lou and I like beer. I very much like Japanese beer, but of the two most prominent beers these days, I very much preferred Kirin over Sapporo. So, for the evening we had to make do with Sapporo, but years later we had our first choice when we had lunch at the Kirin brewery in Yokohama with Jiro and Akiko Matsumoto.

My schedule for the week was pretty busy with discussions and seminars at the Sapporo Medical College, Hokkaido University Faculty of Science, and the Department of Dermatology of Sapporo Medical College. Evenings meant various dinners with the many people I met. At the Hokkaido University Faculty of Science, I had the opportunity for a reunion with Tomoji Aoto, who I had known thirty years earlier in Iowa City when I was a graduate student and he was a post-doc of Professor Witschi.

Somehow during the week, we had the opportunity for some sightseeing in Sapporo. It is a beautiful city, and we saw it at its best for it had been spiffed up a few years earlier for the Winter Olympics that had been held there in 1972. Among the many parks that we saw was one that featured a statue of Professor William S. Clark, a man whose work and influence was reflected throughout the island. The statue depicts Professor Clark standing tall with an arm lifted upward and out and pointing to the future. He is uttering those famous words, "Boys, be ambitious." Professor Clark, professor of agriculture at the University of Massachusetts, was brought to Hokkaido in the mid-1870s. At the time, Hokkaido was relatively underdeveloped, and the Japanese government sought guidance in establishing a strong agricultural system. The land resembles strongly that of our northern tier of states and the northeast, and thus the Japanese chose Professor Clark for advice. We were amazed when we arrived in Hokkaido to see fields and farm buildings much as we knew them at home in Iowa or in New York. I had already traveled considerably in Japan, but here on Hokkaido, we saw silos and barns for the first time. The story of Professor Clark has been a source of amusement for me over subsequent years. I remember returning home from Japan on Japan Airlines, and when I found the stewardesses pleasant and with seemingly good senses of humor, I would ask, "Who said 'Boys, be ambitious.'"? The immediate response was "Professor *Crark*" and a giggle. It seems that all Japanese school kids are taught the story of Professor Clark.

The problem of what we would do over the weekend was solved early in the week when Kowichi Jimbow asked if we would like to spend the weekend with him and his family at Lake Toya, a few hours away by car. Kowichi's wife's uncle owned a vacation home there, and the house was available that weekend.

Of course we agreed, and so on Saturday morning, Lou and I piled into the Jimbow Land Rover, along with several kids, and away we went. Lake Toya is in a zone of small volcanoes, and in fact, one of them, just three or four kilometers away, was erupting. It was a spectacular sight! There were small tremors all evening long, and the pictures were dancing on the walls.

Sleeping was a bit anxious as the tremors continued all night. The next day was beautiful, and I went out to toss a Frisbee with the older kids and to carry the youngest little girl on my shoulders. In the afternoon, we headed back to Sapporo, making a brief stop *en route* to visit an Ainu village and museum. It was fascinating to learn about the Ainus, a proto-Caucasian people who were the first inhabitants of Hokkaido and who fought fierce battles with the Japanese invaders from the south. The next morning, we flew to Tokyo to start the last segment of my Japan visit.

Tokyo

My host in Tokyo was Jiro Matsumoto who did a superb job of organizing the visit even though it was going to be a bit complicated. He got together with another friend of mine from Tom Fitzpatrick's lab, Kiyoshi Toda, to arrange housing accommodations. Toda, of Tokyo Teishin Hospital, which I believe is still the postal workers' hospital, seemed to have a way of getting financial support, and he put his talents to work to provide us an excellent room at the Palace Hotel. I could not have been luckier, for it was my favorite place to stay in Tokyo. It is just across the street from the beautiful Imperial Palace and its lovely gardens. The Palace is surrounded by a moat, complete with swans and water lilies, and around it is a wide, paved pathway that is much used by walkers and joggers. The pathway runs for six kilometers around the palace, and on several visits to Japan I jogged it to provide much-needed exercise.

The schedule that Jiro drew up was perfect for, while I was scheduled to give numerous talks and seminars, there were days off for rest, relaxation, and preparation. Jiro had informed Prince Masahito of my schedule, and since the Prince and Princess Hanako were to be free during the early part of the week, they invited us for formal tea at their home in Shibuya. Lou and I put on our finest clothes (poor Lou had not brought anything fine, however) and basked in the superb royal hospitality. After having tea, we strolled in the lovely garden of their Shibuya home. Of course, this was a memorable experience that is etched in our respective memories. Who would have ever thought that Lou, from a small town in Iowa, and I, the son of Italian immigrants, would ever have had such an experience.

The next day was a little more pedestrian as we went over to the Hiyoshi campus of Keio University to visit with Jiro, Mat, their colleague Sumiko

Negishi, and others. I am not sure what position Sumiko had at that time, but I believe she was associated with Mat. Sumiko is a sweet person, and I enjoyed seeing her again in later years. In the afternoon, we went over to the Tokyo campus of Keio University for me to speak about the developmental physiology of pigment cells. We had a little respite the next day, but that evening we took the train to Kamakura to have dinner with Mat and Yoshiko at the Obika home. It was close to midnight when we returned to Tokyo Central Station and walked back to the Palace Hotel.

When we collected our room key, we were informed that there was a package that had been left for us along with an accompanying note. They were left there by Tadao Hama who knew that we were to be in Kamakura that evening. In his shyness, he delivered the package while we were gone. He had taken the train from Nagoya, dropped off the package, and then quickly returned. What was in the package was even more surprising. It was the new model Olympus camera that Hama had told us about the morning of our departure from Nagoya. Even more astounding, Hama had taken the instruction manual, which was written in Japanese, and word for word had translated it into English! Lou was to return home shortly and left the camera with me since it needed to be declared for customs, and we had no notion of its value. I still had a week or more before my return home, and sometime during this interim, between engagements, seminars, and meetings, I found time to go to the Ginza where there was a superior camera shop that I knew of. In fact, I had planned to go there before my return home in order to buy some extension tubes for my own camera, a Topcon. I knew that someone there could tell me the value of Lou's gifted Olympus. It turned out that the clerk was a soul mate in that he, too, was a Topcon owner. Topcons were not very common in the states, but in Japan there was a cult of proud Topcon owners. When I told him about the Olympus camera and how we acquired it, he quickly remarked that we could not possibly have English instructions for they had only become available that very day. I told him about Hama's translation, and he could hardly believe it. Being the nice person that he was, he went to the stack of newly arrived English versions of the instructions and presented me with a copy.

Much went on this last week, but an important thing for me, on the personal level, was to go to Toda's Tokyo Teishin Hospital to present a seminar. I felt beholden to Toda for his generosity and help. Moreover, I liked him very much and I hope that I did him justice with my presentation on "Animal models of human pigmentary phenomena."

And so my fascinating forty-five-day visit to Japan came to an end. What a wonderful and enriching experience it was. I was honored and privileged to be a Research Fellow of JSPS.

Japan revisited

The very next year, 1980, presented the next opportunity to visit Japan and Sendai again. I needed to be there for the XIth IPCC that Makoto Seiji hosted. I chaired a session, and I had other chores related to my position on the IPCS Council. It was good to see my Sendai friends again and to appreciate the success of the conference. This was a short visit and the last one for another ten years.

The XIVth IPCC was held in Kobe, and again took me to Japan. By this time, I had become an old timer who had been in attendance at IPCCs since the mid-1960s, and so I was asked to give a plenary talk about past conferences. Moreover, this was the first IPCC to follow the one I had hosted in Tucson. In the meantime, I had founded *PCR*, and the IPCS was being replaced by a federation of regional societies, The International Federation of Pigment Cell Societies (IFPCS). Its charter was approved in Kobe, and *PCR* was the official journal of the IFPCS.

After the Kobe meeting, I had the good fortune of being invited by Toshi Fukuzawa to visit his mother and aunt in Nagano. Toshi was from a successful family of farmers. After he left Japan to come to my lab in 1987, following his father's death, his mother took charge of the farm holdings. Soon after the Kobe meeting was over, I took the long and scenic train ride through the mountains to Nagano. Because of its mountain location, there is excellent skiing in the area, and it was the site for the 1998 Winter Olympics. I knew that Nagano was a fairly sizable city, but I was surprised to see such a remarkable mixture of agricultural and urban activities together. The Fukuzawa home and farm buildings were right in town adjacent to rice paddies and city buildings. The visit with Toshi, his mother, and his aunt was fascinating. Their home was sizable and was an amazing mixture of the modern and the old traditional Japanese. We ate at a low table with no chairs and, while my hosts had no trouble with operating on their knees, it was a bit of a chore for me. Nevertheless, it was fun for me even as Toshi was the only other English speaker. We had a fine traditional meal, and at the same time watched the evening news on the large TV that was near the table. I was to sleep in a large space set aside by moveable screens. Of course, there was no bed or mattress; instead I slept on a large futon with pillows and a comforter. The toilet and bath facilities were accessed by an outside corridor. Soon after my arrival, Toshi explained how the toilet worked. It was a quite sophisticated Toto commode operated by several buttons that one pushed depending upon what was required. There was no toilet paper; instead, one pushed one button for a spray wash and another for a stream of dry, warm air to dry with. All of the buttons had Japanese directions, but fortunately Toshi had taped English translations alongside each button. It was altogether a fascinating experience enhanced by the hospitality and friendliness of the Fukuzawa family.

Two years later, in 1992, I was in Japan again to attend the International Symposium on Amphibian Endocrinology where I served as a session chairman and presented an overview lecture. In the intervening two years since my last visit, Toshi had been offered and accepted a position at Keio University joining Jiro and Mat. We had fun as I teased the three of them as comprising the "Tucson Mafia of Keio University." I really enjoyed the symposium since it was so specialized on amphibians and because it was autumn and a beautiful time of year in Japan. When I went to the Hiyoshi campus of Keio University to visit the boys, I was greeted by an aisle of Ginkgo trees in their full fall colors of gold that made the main gate of the university most welcoming.

Another special memory of that 1992 visit was my role as visiting professor at Toho University in Maebashi, way around to the east side of Tokyo Bay. Maebashi seems to be a suburb of Funabashi near the larger city of Chiba which is close to Narita Airport. I was invited to Toho by my good friend Ryozo Fujii, whom I mentioned in reference to the Hokkaido visit years earlier. Ryozo had been a professor at Sapporo University, a state university. In those days, professors at state universities were subject to obligatory retirement at the age of sixty, presumably to make way for younger people moving up the academic ladder. Toho University, on the other hand, was a private university of consequence and free of that age limitation. Thus, Ryozo was appointed professor at Toho, following in the footsteps of an even more eminent professor who had earlier been confronted by the state university age limit. He was Hideshi Kobayashi, one of the pioneers of comparative endocrinology and an esteemed professor at Tokyo University. As I mentioned, he was a dear friend of Aubrey Gorbman and Howard Bern, the two fathers of comparative endocrinology. After having left Tokyo University and serving as director of the famed Misaki Marine Laboratory, Hideshi relocated at Toho University. I believe that it was he who brought both Aubrey and Howard to Toho University as visiting professors. This function was an annual event for Howard, who made this visit to Japan for many years. I was, therefore, very much honored in 1992 to be asked to follow in Howard's footsteps. It was also a real pleasure to see that Ryozo was back on his feet and functioning very well at Toho. His colleague there was Noriko Oshima, a younger woman of considerable abilities and accomplishments. Apparently, my brief stay at Toho was satisfactory, for Ryozo and Noriko asked me to return the following year for a longer appointment and for more teaching activities.

During the ensuing months, we made plans for our return to Japan the following fall of 1993. Lou was to accompany me so that we could make a journey that we had long talked about, but had never realized. We wished to make a pilgrimage to Kyushu, the easternmost of the large islands of Japan. It was a family matter that I knew could be most easily accomplished with the help of one of my friends, Yoshiaki Hori, a noted dermatologist at the Medical College in Fukuoka. Hori had been part of the group that worked with Tom Fitzpatrick in Boston. I wrote to Yoshiaki and related the following story.

Lou's maternal grandparents were Nicholsons, and the eldest of her cousins was Rodney Nicholson, Jr., about fifteen years older than she. She adored Rod, and the feeling was mutual. Both Rod and his father, Rodney, Sr., were fliers, and during WWII, when Rod, Sr., worked in the aircraft industry, Rod, Jr., enlisted in the Army Air Corps. Rod was sent to Waco, Texas, for training and successfully obtained his wings. He had hoped for combat duty, but instead spent most of the war as an instructor. Eventually, he made full Army, and thus the Army was to be his career. After the war was over, he was assigned to a base in Japan on the island of Kyushu. In the meantime, Rod had married and, not long before he was assigned to Japan, he and his lovely wife, Marian, became parents of a little girl, Sandy. Rod went to Fukuoka and was later followed by Marian and the baby Sandy, now eighteen months old. They had just acquired family housing, and Rod went to the base to borrow a jeep to transport their belongings. When Rod arrived at the base, two new fighter planes had arrived and needed to be flight tested. Rod, who would never let the opportunity to fly go by, offered to test fly one of the planes.

The base was near Kitakyushu, a small city not far from Fukuoka. When Rod's plane was over Kitakyushu City, it developed a serious malfunction. Witnesses were convinced that, rather than bail out, Rod stayed with the plane in order to change its trajectory to prevent it from crashing into a school. Of course, Rod perished in the crash. One of the witnesses of the crash was a Mr. Maeda whose little daughter was a student in that school. Mr. Maeda was a teacher (*sensei*) of some sort of history, perhaps of martial arts. He was so taken by this act of heroism that he felt that Lt. Rodney Nicholson had displayed the spirit of the Samurai. He built a shrine at his home in Kitakyushu, and the people of Kitakyushu City erected a small monument to Rod at the site of the crash. Some years later, the Mitsubishi Company, which had a plant in Kitakyushu, wished to build a road that would impact the monument. Out of respect for its significance, the company asked to move the memorial. In doing so, they engaged a noted sculptor to create a new monument to be located in a small park near the crash site. It is slightly elevated on a rise of land, and at the base of the monument, a description of the event and of Rod's heroism are recorded on a bronze plaque in both Japanese and English. When the monument was dedicated, the Mitsubishi Company invited Marian and Sandy and Marian's husband (she had remarried) to the ceremony.

Yoshiaki Hori was so taken by this story that he took it upon himself to go to Kitakyushu City to meet the Maeda family. The old *sensei* had died, but his son and family were living in the house. Yoshiaki graciously wrote a description of all that had transpired for the Fukuoka Medical Society Newsletter. In addition, he asked us to come to Fukuoka late in November to attend and to speak at a dermatological meeting. He arranged for us to visit Kitakyushu City to see the monument and to meet the Maeda family. We arrived on November 20 on the Hikari (one of the bullet trains) and were installed in the superb Hotel Nikko Fukuoka. The next morning, we took a local train to Kitakyusku to meet the Maedas to visit the lovely monument and the original crash site. The Maeda home was very nice as was their traditional Japanese garden. The original shrine that Sensei Maeda had built in his garden was still there and had obviously been well maintained. Notwithstanding the serious nature of the visit to Kitakyushu, we had a pleasant day and returned with a sense of fulfillment and satisfaction.

We looked forward to the evening when we were to be with Yoshiaki and Yukie Hori and other friends that I knew quite well. These included Kiyoshi Toda from Tokyo and Madhu Pathak from Fitzpatrick's lab at Harvard. The occasion was to have a "Fugu" dinner, an extremely coveted and notorious fish quite common in the waters around Kyushu. Fugu is a puffer fish that is extremely toxic, especially in some seasons of the year, because it contains tetrodotoxin, a sodium channel blocker. Only special Japanese chefs know how to prepare it. I do not remember anything special about the meal, but I later realized that it must have cost the Horis a small fortune to host us all. The next day was the dermatological meeting which included my talk. During that time, Yukie and Lou took in some of the cultural offerings of Fukuoka. Our brief stay in Fukuoka was delightful, and Yoshiaki and Yuki became dear friends forever after. We left and moved toward Tokyo with overnight stops in Hiroshima and Kyoto before arriving at the shin-Yokohama station. We were met by Jiro and Akiko and taken to their home, the "Matsumoto Waterbed Ryokan." The waterbed was a relic of Jiro and Akiko's daughter, Mina, who had long since moved to her own apartment in Tokyo. We stayed with the Matsumotos for two nights, long enough to get together with my Keio friends and for all of us to party with the Tucson Mafia. The last stop in Tokyo was for me to speak again at Toda's Teishin Hospital with an overnight stay at our favorite Palace Hotel, again thanks to Toda's hospitality.

We left Tokyo the next day and returned to Maebashi and Toho University for several days as visiting professor. In addition to discussions about research with Ryozo, Noriko, and their associates, what a wonderful experience it was to have one-on-one interactions with some of the undergraduate students. These were particularly happy days for us, and we were sad to leave. Our last memory of Toho was being taken to lunch by the students, so we left under congenial circumstances marred only by the fact that I was deprived of my last chance to have a bottle of Kirin beer with lunch. The restaurant served only American beer! After lunch, the whole gang of us left for the train station with our baggage, and they saw us off as we proceeded to Narita Airport for our flight home.

It would be another six years before the opportunity arose for our penultimate trip to Japan. That was in 1999, and the occasion was the XVIIth IPCC which was held in Nagoya. Lou was with me on that trip, probably because the Matsumotos invited us to visit them again. Moreover, Lou had never been to Nikko in a temperate time of the year, and I wanted to take her there. We went directly to Nagoya where I was a speaker at the opening ceremony. Jiro was also heavily involved in the conference and brought Akiko with him. Thus, while Jiro and I participated in the IPCC, Lou and Akiko took advantage of the accompanying persons program and had a fine tour of Nagoya's points of interest which they supplemented with trips of their own.

Afterwards, the four of us returned to shin-Yokohama and the Matsumoto home for another overnight visit. The next day, we drove to Nikko for an overnight stay that allowed for a good visit of the temples, Lake Chuzenji, and Kegon Falls. During the trip, we learned that the Matsumotos were to celebrate their wedding anniversary during our stay. Of course, I insisted on preparing an Italian dinner to celebrate the occasion. What could we prepare in a Japanese kitchen? As I thought about it, the

Akiko and Jiro Matsumoto at Nikko National Park.

idea of ravioli emerged. On our return from Nikko, we went directly to a local supermarket with Jiro and Akiko. They had flour and eggs at home, so I had the ingredients for the pasta dough. At the cheese counter, we found something that could pass as ricotta and which I could flavor with parsley and other herbs. The tomatoes looked pretty good, so Lou offered to make a pummarola sauce. We thought of *prosciutto e melone* for an antipasto, since we found some sort of cured ham that resembled prosciutto. However, melons were really expensive, upwards of twenty dollars apiece, so we substituted fresh figs. We would start out with *prosciutto e fichi* instead! We had the good luck to find a bottle of an Italian red wine that we knew, *Ecco Domani rosso*, so we were set to go.

In Akiko's typical Japanese kitchen, I found some surface space to roll out the pasta dough that I made, and we cut some rounds into pieces for the ravioli. I mixed a filling of egg, ricotta, and chopped parsley flavored with some grated romano or parmesan cheese that we had found in the market. Lou cooked up the fresh tomato sauce, and in no time we had dinner ready. Altogether, I'd say we did a good job, and the Matsumotos were too polite to say otherwise. For us it was a wonderful visit which, at the time, we thought would be our last. We hoped that we could somehow repay Jiro and Akiko for their kind hospitality, and thankfully we were able to do so the following year. Jiro needed to be at a PASPCR meeting in Texas, so we invited them to come early to be with us in Tucson and Greer, and to visit Canyon de Chelly and Santa Fe before going to College Station, Texas, for the meeting.

22

The Retirement Years

I officially became professor emeritus in 1992 at the age of sixty-three, a pretty young age to retire; however, sixty-three is only a number and not really an indication of the real state of affairs. I chose to retire formally at that time because there was a narrow window of opportunity that would ensure the highest rate possible for my pension. I retired knowing that I could continue my teaching activities as usual at a pay rate of forty-nine percent of my former salary and at the same time receive my pension. My laboratory would continue as before, so it was all a net gain to retire as I did. I loved teaching my basic developmental biology course, so I continued for two more years. I would have given my undergraduate course for even a few more years, but I chose not to for several reasons.

In 1983, when my position was moved from the College of Arts and Sciences on main campus to the Department of Anatomy in the College of Medicine, I was placed somewhat in limbo. The courses I taught were officially liberal arts courses. Moreover, the space I occupied was really under the aegis of Liberal Arts, and specifically the Department of Molecular and Cellular Biology (MCB). This meant that all the equipment in my laboratory that had been purchased through my research grants was on the Department of Anatomy inventory. On the other hand, my courses were liberal arts courses administered through the Department of MCB. Personally, my home department was Anatomy, and thus my boss was the head of the anatomy department. Most of my time was spent on main campus where both my lab and courses were situated. It worked out for the ten years before I retired, but I never had a real sense of belonging to either Anatomy or MCB. I was neither fish nor fowl, a status not at all helped by an anatomy chairman who was rather remote and obviously cut from the same cloth as were old-time heads of departments. In those days, the headship was a reward, an indication of having arrived, not as in the more modern view where the chairman is a person of respect who makes sacrifices for his department in order to help each member do his or her work more easily. Fortunately, I related well and enjoyed associating with some of the young, new people in the department.

On main campus, course curricula were beginning to change, trying to keep pace with the rapid developments in cell biology, genetics, ecology, and biochemistry. Moreover, teaching methods were changing with many courses being team taught, not so much to increase the efficiency of teaching, but to allow faculty more free time for their own research endeavors. As an old mossback, my view was that, as an academic professional, my job was both to disseminate knowledge and to create it. With respect to the former, it was a pleasure to meet and know young people who would move on to greater service and creativity than I. With the latter, the sense of accomplishment in a new discovery was hard to beat. It provided a sense of creativity and satisfaction. Moreover, I had a certain amount of control over my research activities, unlike the teaching of my developmental biology course that had become more and more linked with what was taking place in MCB. I decided that it would be best if I stopped teaching this developmental biology course that I had created thirty-five years earlier, and so I did, and the course expired.

My new emeritus state in no way slowed me down. The lab was still working on MIF with the help of graduate students and with Preminda Samaraweera who had not yet left for New York. Toshi Fukuzawa had already returned to Japan with most of the project left in his hands. I was still very busy as editor of *PCR* and organizational affairs related to the pigmentary world. There were many manuscripts underway or in press that were generated in the lab by me or my students. Francesco Mangano was still working in the lab as he awaited acceptance to medical school, and Warren Johnson was in the final phases of his PhD work. A new PhD student, J Newton, was just beginning his work, and the plan was for him to approach the MIF problem in a different way.

Most of the work that we were doing concerning MIF (and MSF) was based upon using media conditioned by exposure to dorsal or ventral skin. We were thus dealing with a putative factor that Toshi and Preminda were working hard to characterize. Even with our having generated an MIF monoclonal antibody, characterization was not completed, and this task was left for Toshi to pursue after his return to Japan. I felt that searching for the identity of MIF by conventional means of extraction was like looking for a needle in a haystack. I started thinking about approaching the problem through the means of molecular genetics. My passion in this regard was enhanced when we found, through collaborative work with Spanish colleagues, that MIF seems to block the MSH receptor, much as does the small protein, agouti signal protein (ASP), in the mouse. ASP is encoded by the agouti locus, so if we could find an "MIF locus," we might be able to obtain an "MIF protein." J Newton was one of the authors of our paper on the blockage of the MSH receptor, so it seemed natural for him to search out the MIF protein through the means of molecular genetics. I was able to obtain an NSF grant to carry out such a study, so we sent J to Greg Barsh's lab at Stanford to undertake the project for his PhD dissertation. Barsh's lab had the experience and facilities for this work and, in fact, had worked on the agouti gene expression. With an eye toward finishing a dissertation as soon as possible, J turned his attention to a related and simpler project, a study of two genes, agouti and extension, and their role in pigmentation patterns in the domestic dog. Thus, the search for an MIF protein was never done. The MIF problem was too esoteric a problem for J or others in the Barsh lab, and it thus remains the last item in the string of my unfinished business. It is the uncompleted project that I most regret. Its solution requires expertise that I do not have, and younger people will not take it up because the realm of comparative biology has become *passé* and extravagant in these modern times.

Aside from the MIF work, I was still involved with several other projects that required a lot of time. As I look back at my publication activity in 1992 and 1993, I find that these were the most prolific in my career. Perhaps these represent a great way to go out, but I still had other things to do, many of these had to do with the field of pigment cell biology. Before I recount endeavors in this regard, let me tell about a pleasureful interlude in 1994–1995 that concerned the presence of Paolo Chieffi in my lab and in our lives.

Paolo is the youngest of Giovanni and Annamaria's six children. He had grown up to follow his father's footsteps by going to medical school and then to undertake a career in basic research. His father felt that Paolo should be better grounded in developmental biology. I tried to explain to Giovanni that the rapid development in new methods of research had bypassed me, and that Paolo needed to be in a modern, active lab. He would have none of it and insisted that Paolo should come to me to be exposed to the basics of developmental biology and at the same time sharpen his use of English. From the personal standpoint, that was fine with Lou and me, because we had known and loved Paolo since he was a little boy.

We both remember well one Sunday morning in Naples during Lou's mother's visit. Giovanni had volunteered to lead Hazel and us on a walk through the old historic area of Naples. The reason for this particular excursion was that Sergio, Giovanni's ten year-old son, had a passion for old historic Naples, and it was he, actually, who was our guide. Giovanni also had young Paolo with him. We believe that Paolo must have been about three years old at the time. At any rate, it had rained earlier in the week, and there were still puddles. As we passed along the cloister walls of the famous Santa Chiara church, somehow some muddy water splashed up to soil Paolo's pristine white suit. Giovanni scolded Paolo mildly and, of course, the little boy burst out in tears, and we all just melted. Now, some twenty plus years later, Paolo was to be our guest in Tucson.

After we got Paolo settled into an apartment and my lab, I decided that the best help I could offer him was to see that he be introduced to a modern, active lab. I knew, but not well, Parker Antin, a young developmental biologist in the College of Agriculture. We walked to a nearby building to meet Parker, who welcomed Paolo into his group. The union was most successful for Paolo dug in and worked hard; moreover, he was extremely well liked. Soon they published some of the work that Paolo did. On weekends, Lou and I entertained Paolo, and we enjoyed showing him our favorite places in this part of the state. Paolo was, of course, impressed, for, having grown up in a bustling city like Naples, this was another world. One Sunday morning, we drove to the Southwestern Research Station in Portal where my former student, Wade Sherbrooke, was director. SWRS is located on the east side of the Chiricahua Mountains on Cave Creek, which is, I believe, the most beautiful area in Southern Arizona.

Since Paolo was young and vigorous, I offered one Saturday morning to share with him my passion for hiking. I suggested a hike because it was a way to appreciate the unique history of our area. One of the places that Charlie Thornton and I had frequently hiked was along Sutherland Wash on the west side of the Catalina Mountains, just south of the town of Catalina.

We liked to follow some primitive trails here, not only because it was scenic, but because we often found remains of Indian tribes, probably Hohokam, that had inhabited the area. It was rich in pot sherds and many of these seemed different from sherds in other sites in the vicinity. These were probably the remains of painted pots, for one could clearly see colored designs on some of the larger pieces. Often we would find sherds that had a curved edge and, more than likely, they had broken off from the lip of a vessel. Often, as we walked along, we would come upon large flat rocks that were embedded in the soil. Some of these were probable metate sites, for there were deep holes in them. Very likely, Indian women would use these depressions to process mesquite beans into flour by grinding them with a mano made from a smooth stone. This arrangement was the American Indian version of a mortar and pestle. In our wanderings, I had never found a mano, and I do not know if Charlie, my old hiking pal, ever did.

It was a beautiful morning as Paolo and I hiked along for several miles. I usually led, with Paolo right behind. Somewhere along the trail, as we crossed a small stream bed, Paolo said, *"Joe, questo che è?"* I turned, and there in his hand was a beautiful mano. When I explained what he had found, he offered it to me. I said, "No, Paolo, you have found a wonderful souvenir of your trip to Arizona."

Paolo returned home earlier than he and we expected. A faculty position had opened up in one of the life science branches of the University of Naples, and there was a *concorso* (competition) for that position. Given Giovanni's powerful place at the university and in Italian science, there was a good possibility that Paolo could win the *concorso*, and so he did. We see Paolo frequently when we return to Naples, and it is always such a warm reunion. He has made a real success of his career, and his research goes forward building on his training in Parker Antin's lab.

Since the period that Paolo was in his lab, Parker moved to our department. We often spoke about Paolo, and I learned the origin of Parker's name. I had never paid much attention to his last name except that I thought it unusual. Parker's parents came to Tucson to visit and, as we talked, I realized that Antin was a modification of the Italian family name, Antinozzi. When Parker went to Italy to visit, Paolo took him to the Antinozzi ancestral town.

The "Blues" mini-symposium

When I actually stopped teaching in the 1994-1995 academic year, I had more time to devote to other matters. In fact, my encumbrances decreased even more because 1994 was the end of my tenure as editor of *PCR*. During that seven-year period when many manuscripts dealing with pigment cells passed through my hands, I realized just how great a divide separated the two poles of pigmentation researchers, the basic biologists at one end and the medically oriented workers at the other. Myron Gordon's original intent of bringing together investigators from the two ends of the spectrum was successful, but there nevertheless remained a rift that the less open-minded members of our societies scarcely looked across. On the other hand, many of us did profit from the knowledge exposed by the opposing sides.

As the time for the XVIth IPCC approached, I decided to try to point out one of the little recognized commonalities that exist among our broad family of investigators. The IPCC was to be held in Anaheim under the chairmanship of Frank Meyskens who had formerly been a member of our own Cancer Center. During his Tucson years, Frank had worked hard at bringing together researchers from all aspects of pigmentation and cancer. In fact, he had organized an annual retreat which many of us attended to share our research. I suspected that Frank would welcome my suggestion of holding a mini-symposium on blue coloration, a subject of interest to both comparative biologists and dermatologists. It was a subject that had interested me for years, and, when I worked on the blue spots of torpedo in 1971, I realized that these spots were essentially like the blue nevi of human skin. In recent years, there had been numerous papers published on the blue coloration of fishes, and one of Ryozo Fujii's students had just demonstrated the presence of blue pigment cells (cyanophores), containing a true blue pigment, in the skin of a coral reef fish. Accordingly, I contacted Frank with the proposal of organizing a mini-symposium with the help of two dermatologist friends, Yoshiaki Hori from Fukuoka and Jean Bolognia from Yale. The detailed program that we proposed met with the approval of the organizing committee, and we were put on the conference program.

A major presentation was by Craig Bohren, an atmospheric physicist from Penn State, who told us about light scatter and how it was the phenomenon behind Tyndall or Rayleigh scatter. The structural blue colors of insects, fishes, reptiles, birds, and mammals (including humans) are products of Rayleigh scatter. These structural blues are presented by skin, feathers, eyes, and chromatophores, and thus provide a common theme. We found colleagues of note to speak about various animal classes and about pathologies, and this attracted a significant audience. Listeners must have found the presentations interesting because they stayed, and by the end of the conference, many in attendance approached us with the compliment that it was the best session they had attended. Be that as it may, we were pleased with what we had done, and the thought arose that we might solicit chapters from the various presenters and publish a symposium volume. Yoshiaki Hori and I were to be editors of the proceedings. When we all returned home, the end-of-the-year distractions took over. By the time we got back to the symposium volume, Yoshiaki became ill, afflicted by an aggressive cancer that took his life. The symposium volume was shelved, only to be resurrected in another form several years later.

The XVIth IPCC was a grand success, not only because of the science presented, but because it was also a great social event. We were in the shadow of Disneyland, and I'm sure that everyone at the conference must have found time for a visit there. Lou and I went over with Jiro and Akiko Matsumoto. Even in our advancing years, we had great fun, and the young people there had nothing on us! For me, the IPCC had special significance, for it was the first such international conference since my retirement a few years earlier. Many of my Japanese friends and former associates, led by Toshi Fukuzawa, decided that they wanted to hold a party to celebrate my retirement. Toshi organized a small banquet, and we all had a great time together. Of course, I was touched to be so honored, and I was moved even more when they presented me with an impressive gift, a Nikon F70 state-of-the-art camera. It must have been quite expensive, and I guess everyone pitched in to pay for it. Such friends!

It is ironic that my retirement made little impression at the University of Arizona. Except for the various clerical procedures that needed to be done, little notice was taken. Fortunately, I was not totally ignored for, in 1995, my retirement was feted at the University of Kansas. At that time,

Sally Mason, then Dean of Liberal Arts at Kansas, was hosting the PASPCR meeting. She organized a party to recognize my retirement, and I was so pleased that among those in attendance were not just pigmentation people, but friends and former students who made the trip to Lawrence just to be at the party. I have a wonderful souvenir of the event in the form of an original print that Sally presented to me. It hangs on the wall of our family room for us to enjoy every day.

During the remaining years of the millennium, I seemed to be involved in a lot of writing. One significant and long effort was a chapter to be published in two volumes covering pretty much every aspect of pigmentation from the basic science end to the clinical. This work, *The Pigmentary System*, was edited by friends and colleagues from the pigmentary world, Jim Nordlund, Ray Boissy, Vince Hearing, Dick King, Bill Oetting, and Jean Paul Ortonne. The first edition was published in 1998, and my early chapter, "Comparative anatomy and physiology of pigment cells in non-mammalian tissues," was the longest. This authoritative tome was so successful that, a few years later, a second edition was called for. This was a more streamlined version that excluded some of the chapters from the first edition. I was fortunate enough to have my chapter included. In the years since the first edition, there were many new developments, especially in the area of molecular genetics, and I knew that I needed help if I were to produce a first rate revision. Accordingly, I asked Jiro Matsumoto to be co-author. He agreed, and we thus produced an excellent chapter for the second edition that was published in 2006.

Another publication that I became involved with during this period was one that I did not seek. Nevertheless, in the end it all worked out. It seems that after the mini-symposium on blue pigmentation, and we all had returned home, Ryozo Fujii began to prepare his chapter in anticipation of a symposium volume. With Yoshiaki Hori's death, he put the project aside. A few years later, Ryozo himself was diagnosed with prostate cancer. The prognosis was not encouraging, so he commenced a flurry of activity with the hope of publishing several partially prepared manuscripts. Even when he was in hospice, he asked his daughter, Sana Harada, to help him complete these manuscripts. One of the papers was one he outlined for the blues symposium publication, and it was in an envelope together with potential illustrations and figures. When he realized that he could never complete the work from his hospice bed, he asked his daughter to send it to me to finish. He expired soon thereafter, and I was left with the manuscript to complete for publication. I agonized over this project for several years, knowing that, while I had some knowledge of the subject, I was not competent enough to do Ryozo's work and write a manuscript that would pass the scrutiny of referees of a specialty journal. I might have forgotten about the whole business, but eventually my conscience prevailed, and I came up with a plan that might serve many purposes.

My thought was that, while I may not have been up to writing an original detailed paper based on Ryozo's experiments and observations, I could certainly do so in a more general way. Moreover, I could do the same for the other presentations given at the blues mini-symposium and thus incorporate it all together in the form of a review article. By doing so, the blues mini-symposium could be rescued from the bin of unfinished business at the same time that I did my duty to Ryozo Fujii. The more I thought about this plan, the more enthusiastic I became, especially since, at the end of the 1990s and the first few years of the 2000s, a series of excellent papers had been published about the anatomy, physics, and evolution of avian structural colors. This was the work of Richard Prum, a young evolutionary biologist from Yale University. I wrote the outlines of a review article, "On the blue coloration of vertebrates" and asked Phil Fernandez to be co-author along with Ryozo Fujii as posthumous co-author. Phil had been one of the participants in the mini-symposium, and although I knew he was busy with his teaching at Glendale Community College, I hoped that he would be able to spare a little time with the manuscript. Thus, a pretty darn good review article was generated and sent to Colin Goding, the then editor of *PCR*. He liked it and published it in 2007. Furthermore, he accepted an elegant photo of a blue-footed booby taken by one of Phil's friends to be used on the cover of the issue in which our review appeared. Thus, my conscience was eased, and Sana Harada was most pleased to have her father so recognized. We noted as a footnote that the essence of the review was derived from the mini-symposium, and thus this bit of unfinished business was finished.

The review on blue coloration was the last paper that I have written, and I do not anticipate any in the future. Scientific progress has sped along so rapidly that, at this point, I would have nothing to write about except perhaps something of historical significance. Perhaps an exception would be something in the realm of natural history, and nothing serious at that. For about the last ten years, Lou and I have spent the better part of the five summer months at our cabin at Greer, Arizona. We spend as much time as possible out of doors, and this means that I am constantly aware of nature. For the almost forty years that we have owned our cabin, I have kept a kind of log and, just before retiring for the day, I jot down some of the day's significant events. This sometimes became amusing, for I was adamant about writing, even when being very sleepy. As a result, some of what I recorded was done in my sleep, and it was amazing what I wrote down. One such episode has become quite a joke between us and brought a new word to our vocabulary. We had been out one evening, and as we approached our cabin up in the woods, we passed through a little meadow. Sometimes a small herd of mule deer does feeds there in the dark, as they were that night. The deer have become pretty tame, and they essentially ignore us. The next morning as I read the log entry from the previous evening, I found that we had seen fourteen Katchinas feeding in the meadow! Now, when we return home in the dark, we always keep our eyes open for Katchinas!

Aside from these few minor slip-ups, the log has been a great source of information. We can look back to see what the temperature or rainfall was a previous season, or what we served for dinner the last time neighbors came for dinner. Perhaps of more importance, I have recorded the comings and goings of birds and animals. We have a small deck on the east side of the cabin where we often sit, weather permitting. For the length of the season, from a day or two after our arrival to just before our departure, I hang several hummingbird feeders. These can be seen easily from our glass-enclosed porch, so even when we are inside we can observe the feeders. Of course, given my interests and knowledge, I know which species we have, when they come, when they leave, and so on. Many of our friends and neighbors at Greer feed hummingbirds, and we often have conversations about the state of affairs of our hummingbirds and about how much sugar they consume. I have become a bit of a know-it-all when it comes to hummingbirds, so I often field questions about them from our friends. About three or four years ago, I sat down and wrote a short piece on hummingbirds for my friends and neighbors. It was in the form of a letter in which I told about what species we have, their description, when and where they come from and where they go, when they nest, and so forth. The letter became popular, and the Greer Civic Association printed up hundreds of copies for distribution along with other pieces of information available to tourists and visitors. The Arizona Game and Fish Department became interested in my hummingbird letter and asked me to adapt it for publication in their magazine, *Arizona Wildlife Views*. They provided beautiful color photographs of the hummingbird species described, and a very handsome article evolved and was published in the July-August 2010 issue. That was my last publication.

Order of the Sacred Treasure, Gold Rays with Neck Ribbon

One of the highlights of my career, and certainly of my retirement years, I owe to Japan and my Japanese friends. It was announced to the news media on November 3, 2001, that I was to be decorated by the Ministry of Education, Science and Culture of the Japanese government and awarded the Order of the Sacred Treasure, Gold Rays with Neck Ribbon, for my service to Japanese education and science. News of this honor came like a bolt out of the blue on October 31, just a few days before the award was to be presented. I learned about the honor from Jiro Matsumoto who was my sponsor. I should have had some inkling about the honor, but I was just not perceptive enough. In those years, I was in frequent email contact with Jiro, and a year or more before I learned of my decoration, Jiro had asked me to send him my curriculum vitae and some other documents. I was too dense to realize that something was in the works. When he sent me news of the award, he indicated that I could either attend the award ceremony and have an audience with the Emperor, or that the award could be conferred upon me by the Japanese consulate in Los Angeles. It was obvious that Jiro and Akiko wanted us to come to Japan and to stay with them in Yokohama. We decided to accept their invitation, and Lou, who had recently retired from the travel business, called to get us flight reservations. In retrospect, if we hadn't accepted the Matsumoto invitation, I believe we would have hurt their feelings terribly.

Jiro had nominated me to the Ministry of Education, and this entailed a lot of work, for my credentials were subjected to intense scrutiny. Awards to persons from overseas required a three-step screening process following nomination by a person in a prestigious position in an important Japanese organization. In my case, I was nominated by Jiro Matsumoto, then president of the Japanese Society for Pigment Cell Research (JSPCR). Since JSPCR was in its infancy and had not yet achieved wide recognition in Japan, my nomination was given additional support by a long-time friend, Professor Sakae Kikuyama, president of the Japanese Society for Comparative Endocrinology, and by my former post-doctoral associate, Professor Hiroyuki Ide, secretary of the Japanese Society for Developmental Biology. The nomination was based upon my long research collaboration with Japanese scientists, my support of seven Japanese post-doctoral fellows in my laboratory between 1963 and 1991, my contribution to the growth of pigment cell biology in Japan, and my own personal scientific record.

We left for Tokyo on November 6, feeling a bit weird at the airport because, just a few weeks earlier, on September 19, we had returned home from Ireland where we had been at the time of the September 11, 2001, tragedies in New York City, Washington D.C., and Pennsylvania. Now we were *en route* to Tokyo. The trip to Narita and then Yokohama went smoothly, and Jiro and Akiko were waiting for us. November 9 was the exciting awards day. There was so much going on and, as I attempt to describe the events, I am grateful that Lou kept a diary that has helped me remember the details. I took a navy blue suit with us and a red and gray tie, but that was apparently not formal enough. I should have at least had a black and silver striped tie. After some discussion, it was decided that the tie was all right for a foreigner, but a Japanese man should wear a more formal tie.

When we arrived at the National Theater where the ceremony was to be held, I noted that a few Japanese men were dressed more or less as I; however, it seemed that more than half of the men wore morning clothes. The awardees were 98 percent men. Many, many of the wives wore kimono, and the rest wore suits or dresses. We stood in line to register and get credentials. We received our bus and seat assignments, and we each received a rosette and ribbon. As an awardee, I was given a white rosette with a red and white ribbon, while Lou as a spouse received a white and red rosette and ribbon. Jiro was given a white and yellow rosette that I suppose signified that he was our interpreter. We were assigned specific seats in the auditorium. Since I was a foreigner, Lou was seated with me. Japanese wives were seated in the balcony. The auditorium was huge, and it needed to be considering that there were a large number of awardees. Lou remarked that the awardees seemed elderly, and Jiro explained that the Japanese needed to be seventy or older and foreigners at least sixty-five.

The ceremony began at noon. On the stage were the master of ceremonies, dignitaries from the Ministry of Education, Science and Culture, and twelve representatives of each award level or category. The Japanese flag hung center stage behind the seated dignitaries and tables. The presentation was of short duration. First, everyone rose and sang the national anthem—very beautiful with all those men's voices. There were a few welcoming remarks followed by an introduction and presentation of each award category. Each representative rose, bowed to the flag, bowed to the audience, bowed to the dignitaries, then bowed to the Associate Minister, a woman in a beautiful purple suit. She read a scroll, presented it to the representative who then reversed the order of his bows and took his seat. The Minister made a few congratulatory remarks, and then it was time to distribute awards. Row by row, the awardees filed forward to the front of the auditorium where, number by number, each awardee was presented the proper decoration and a tube containing a scroll.

My decoration consists of a medallion in the form of four sets of five rays at the cardinal points radiating from a circle of red gems. The medallion is suspended by a silver and gold moiré ribbon. Among the various documents I received, all written in Japanese, was a sheet in English describing how the decoration is to be worn in public. We exited through a side door, and those of us with neck ribbons were aided in putting them on before we stepped out into the lobby to join the milling crowd of awardees, friends, and relatives, each with a camera in hand. There was not too much time to waste for those of us who were to have an audience with the Emperor. We needed to buy sandwiches to eat on the bus, and more importantly, we needed to heed the advice that was announced from the stage before we received our awards. We were told that we should look to our toilet needs before departing the theater because there were no toilet facilities available at the Imperial Palace. Imagine what a rush there was for the restroom by all of us prostate-challenged seventy-year-olds. I couldn't help but laugh at the comedy of it all-a bunch of old guys all dressed up for a very serious event, stopped in their tracks by a basic human necessity!

Lou and Joe en route to an audience with Emperor Akihito, November 2001.

There were thirty-two buses, but with the aid of a map given us earlier, we found our bus easily. It was a long time before the buses pulled out, and then it took a long time to go the short distance to the Imperial Palace because of the heavy traffic. Finally, three by three, the buses pulled under an awning of the Audience Hall, and we descended.

We went inside and ascended a giant staircase before entering a long, elegant room. Here we were lined up in rows of four, and across an aisle of about eight feet in width, spouses were also lined up in rows of four. I had a good laugh at the ranks of women facing us, because there was Lou, the only occidental among them, standing head and shoulders above the other spouses. I, too, was much taller than most of my fellow awardees. I should point out that not all the awardees were men. On our bus, of the nineteen awardees, two were women. After a bit, a door opened into the center of the room, and Emperor Akihito entered. He walked between the two groups and stopped to deliver some words of welcome. Afterwards, he walked the length of the ranks, bowing from time to time in the typical Japanese way. As he stopped to bow near my location, I couldn't help smiling to myself with the thought, "Hello, Emperor Akihito. I know your younger brother, Masahito." After passing the length of the two groups, the Emperor returned to the door he entered through and made his exit. The audience was ended. We returned as we had entered, and as we gathered under the portico to await our respective buses, each bus group was organized for a photo. There was a row of folding chairs for the awardees, and the accompanying persons stood behind the honorees. Fortunately, the rain had stopped, since the photographer had to stand in the drive to get us all in the photograph. A few weeks later, after we returned home, I received a copy of the photograph. As I studied it, I realized that I was the only one of the men with a neck ribbon. The others had a lapel emblem of some sort. Of the two women, one also had a neck ribbon

The next day, November 10, was also very special. Jiro, in his usual considerate way, had arranged a celebratory dinner for me at the Grand Palace Hotel in Tokyo. There were ten around the table: Jiro and Akiko, Masumi Yasutomi who had come up from Nagoya, Mat Obika, Hideshi Kobayashi, Sakae Kikuyama from Waseda University, Hiroyuki Ide from Tohoku University in Sendai, Toshi Fukuzawa, Lou, and me. All were former post-doctorals of mine except Hideshi and Sakae. The only of my post-doctorals not there were Katsutoshi Imai who is deceased and Tetsuro Iga who lived too far away in Matsue to make the trip.

Hideshi Kobayashi was one of the fathers of comparative endocrinology and a contemporary and dear friend of Howard Bern and Aubrey Gorbman. He was an especially good friend and drinking buddy of Howard's. My friendship with Hideshi is interesting, enhanced by the fact that we would bump into one another in unexpected places. When I taught a summer course in comparative endocrinology at Washington State in Pullman during the summer of 1965, to my surprise I found Hideshi in the office next to mine doing pinealectomies on white-crowned sparrows. A few years later, when we were in Naples at the *Stazione* Zoologica, as I was leaving the building for lunch, I found Hideshi at the front door hoping to find the Director. He had come to Naples, unannounced, from Rome with the hope of having a visit of the *Stazione*. He had just been appointed director of the famous Japanese Marine Biology Station at Misaki, and he hoped to get some tips from the *Stazione*. I knew that there was no one in authority around at the time. Moreover, it was lunchtime and no one would be back for hours, so I took him to a nearby *salumeria*. We had them make each of us a panino, and we returned to the terrace of my lab to eat and visit while we viewed Capri across the bay. A few years later when I happened to be in Japan, I visited him at Misaki and, like Howard and him, we had a few beers together.

Sakae Kikuyama was a contemporary of mine whom I saw frequently at endocrinology conferences in many places. We were both jocks and sometimes jogged together. Sakae was a well-known rugby player, and I used to tease him about being from Waseda University. Waseda and Keio are like the Harvard and Yale of Japan and are most competitive.

The dinner was especially nice, and to have had these many friends and associates together was a moving experience, especially when each person spoke and told about Tucson experiences. We were brought up to date on their families, and that was great. Toshi was my last post-doctoral, and we knew him well since he had spent four years with us in Tucson. After the dinner, we met Toshi's wife, Cheiko, and their two little daughters, Rana and Hyla, beautiful little girls. I had met Cheiko, but had never seen the girls. We wanted to meet them since Rana, who liked to draw, frequently made drawings of us that Toshi sent. It was sad to say goodbye, for we sensed that this would be our last trip to Japan to see these dear friends.

The next day we embarked on a short trip to the Izu peninsula southwest of the Tokyo-Yokohama area. We four have been good traveling companions both in the USA and Japan, so a little getaway seemed like a good idea to cap these eventful and remarkable few days. We went by train to Shimoda near the end of the peninsula. The train ride provided wonderful views of Mt. Fujii as we passed through Odawara on Sagami Bay. At this point, we were very near Hakone, a place I remember well from my 1961 visit where we comparative endocrinologists had an orgy of photography. Shimoda was a lively resort city that I was happy to leave after we picked up our large and luxurious rental car.

We drove across the tip of the peninsula to the Suruga Bay side and descended to the tiny village of Ihama. Here, right on the coast, we found our delightful Pension Costa del Sol. We were the only guests, and the owners, a man and wife, made us feel at home immediately. They were obviously people of eclectic tastes who had traveled. Hence the Spanish name, Costa del Sol, and a menu that reflected Spanish tastes. The hotel was like an art gallery filled with the work of a Venetian sculptor, Livio Di

Marchi. In their travels, the owners met this creative sculptor and commissioned him to come to Japan to design and build their bath house, gate, and fence with imaginative ingenuity. Di Marchi's material was wood, and he carved amazing things such as wooden handbags, curtains, an overcoat and hat, shoes, and many other items. Across the room in the bath house was a clothesline from which hung a wood bra, pair of shorts, and bikini panties! The next morning after a good breakfast, we did some touristic things, such as a visit to a monkey park and a gold mine tour, before heading back to Shimoda to turn in the car and get the train back to Yokohama. The visit with the Matsumotos ended all too soon. We were sad to leave them, knowing that we would not be back to Yokohama. There was no way to thank them enough for their hospitality and for all that they did. Fortunately, we have had the pleasure of their being with us in our home in Tucson since our last memorable visit to Japan.

23

Some Passions

I use the title "Passions" with some reservation since it is a word with different meanings and different degrees. Moreover, everyone has passions to some measure or other. In my case, a love of the outdoors and all that it connotes has been pervasive and has guided many of my activities, even subliminally. I am not unique in this regard, but the expressions of our passions are a personal signature, and so I record them.

As I look back on my life, I realize that one of my earliest passions was to be active and to do hard physical labor. I remember when I was still a boy and spent part of the summer with my grandmother in Victor, one of the things I really looked forward to was going to work on the nearby farm of an Italian bachelor farmer, Ciccio Massino. My uncle Gene and I went to help him with the wheat harvest. In those days, farm equipment was a far cry from that of today. He had a team of horses in lieu of a tractor, and as I recall, they pulled a mower/binder that left shocks of wheat in its wake. For a dollar a day, my job was to stack the shocks of wheat. What a joy it was to do this work and to sweat under the hot sun. Loading mown hay onto the hay wagon and unloading it in the hayloft was an equal pleasure. The fact that I was paid a dollar a day was nice, but what I looked forward to was the work itself. I suppose that my high school job on the NYCRR gave me pleasure both because of the hard work and the considerable wage I earned. This propensity for hard work was put to good use in later life when we had homes of our own, both in Tucson and in Greer that needed to be worked on and maintained.

Handball and Bear Down Gym

As I related earlier, when we first arrived in Tucson, I enjoyed playing basketball in Bear Down Gymnasium. This venerable monument to University of Arizona athletics holds many memories for me. Not only did it provide a venue for many pick-up basketball games, but was a source of entertainment for both Lou and me as we attended every home basketball game. As I recall, admission to the games was free for students, faculty, and spouses. At that time, Fred Enke was head coach, and even though basketball had not achieved the national stature that it now enjoys, Arizona was a stellar member of the old Border Conference. In 1961, the University of Arizona became a founding member of the Western Athletic Conference (WAC). In that same year, Fred Enke retired as basketball coach and was replaced by Bruce Larson. I knew Bruce pretty well, and when a vacancy arose for head time keeper for University of Arizona games, he asked me to fill the position. Since Lou and I went to all the games anyway, I accepted.

For the most part it was fun to do, but the job required much more concentration than just being a spectator. With the small size of the gymnasium, the scorers' table was practically on the court, and the first row of seats was right behind our backs. By the time I became head time keeper, Fred Enke had passed away. His widow and several other elderly widows of retired faculty occupied the seats right behind us.

I served as head time keeper for several years, long enough to provide an array of interesting experiences. Some of these involved very close games when it was imperative that my finger be on the time clock ready to fire on the split second. I will never forget one of these experiences. We were playing New Mexico, and with a second or two left, Arizona was behind by a point. My eyes were fixed on the referee and when his arm shot up, I stopped the clock. There was one second left. The referee had stopped play because of a change of possession, and Arizona was awarded the ball on the sideline at the Arizona end of the floor. The ball was in-bounded to an Arizona player, Albert Johnson, I believe, who scored as time expired, and thus Arizona won the game! I don't remember any repercussions from the New Mexico coaches, but the next day I took a terrible razzing from my friends. During construction of the new McKale Center, I decided that I had had enough of the stress of time keeping. I gave up the job and reverted back to be an ordinary spectator.

During the spring of my first year at Arizona, I began to play handball in the outdoor courts, just outside the back door of Bear Down gym. I played primarily with Dave Snyder, the tennis coach, and with some of the older veteran football players. The latter had been in the military service where they learned to play. We were all pretty competitive, but we played only at a low level. Handball is not an easy sport. It takes a bit of time to develop enough skill to really enjoy the game, unlike racquetball that can be pleasureful from the get-go. Adjacent to the outdoor courts was the University of Arizona baseball diamond and attendant stands. We often played at noontime, just before the start of baseball games. On baseball game days, we were often joined by one of the umpires, Carlos Carillo. He would arrive early enough to play a little handball before his umpiring duties. Carlos had been playing handball for years with a group of his friends at the downtown YMCA. They played every Tuesday and Thursday, and Carlos invited me to join them, and so I did, forsaking the primitive outdoor courts.

Carlos' group of players was mostly Hispanic, meaning that at some time their ancestors were Mexican, mostly from Sonora. I had a lot of fun with these men who were just regular guys. Carlos was a brakeman on the Southern Pacific Railway, another was an electrician, another two were restaurateurs, and still another was a fireman. At that time, I was pretty naive about the Southwest and that those of Mexican heritage were discriminated against by many Anglos. I really couldn't understand this, having grown up in a cosmopolitan city with many ethnic enclaves. In any case, they were my friends, and I enjoyed playing with them and sharing a beer or two afterwards. In retrospect, my one regret about this association is that I did not work at trying to become fluent in Spanish.

As I became more proficient at the game, I began to play at a higher level and with better players. In time, handball took off in Tucson, leading to the emergence of the Tucson Athletic Club (TAC) devoted primarily to handball and racquetball. I became a charter member. At about the same time of the construction of TAC, the national handball organization centered in Chicago began to collapse. Thanks to the efforts of Carl Porter, one of our local players, the US Handball Association experienced a rebirth in Tucson, and thus our town became the handball capital of the world.

Many national tournaments have been held in Tucson and some of the top players in the country relocated here. For many years thereafter, I played handball at TAC at least three times a week, and from time to time I played in the nationals.

When I was on sabbatical leave in Paris in 1964, I looked for a possibility to play. At that time, Supreme Headquarters Allied Powers in Europe (SHAPE) was located at Versailles, just outside of Paris. I was pretty certain that NATO, a part of the SHAPE umbrella, would have athletic facilities, and certainly there would be handball courts. I called NATO to check on this possibility and was referred to the athletic director, Barney Gill. I do not remember his rank; it was either Captain or Major, but in any case he was a West Pointer and could well understand my desire to play handball. The fact that I was a Fulbright Scholar probably helped in his willingness to get me set up to play. The means to get to Versailles and to nearby Camp Veluso was another matter. It ended up that to get there from the lab meant taking the Metro to the Etoile (Arc de Triomphe) where a military bus left for headquarters at Versailles every hour. At Versailles, I needed to take another bus to Camp Veluso and the handball and squash courts. Coming and going to Camp Veluso was a time-consuming affair, but when we finally had our VW, the trip took not much more than twenty-five minutes. It is interesting that the person I played most often was Captain George (Pete) Peterson. Years later, I met him again in Tucson. After his tour in Europe, he returned to the states and was stationed in California and then Arizona. He became a member of TAC and played handball almost until his death a year ago.

While playing handball at the NATO facility in Paris was accomplished with ease, this was not the case for my next sabbatical in Naples in 1970-1971. Again, I was granted use of NATO facilities in the Naples area, but when I started playing handball there, I was stymied by a colonel who was in charge of the gym. There was only one good handball player there, a Captain Brunetti from San Francisco. He welcomed my presence on the scene, for at last he had someone to play with. We played one time until the colonel came upon us and asked who I was. The fact that I was a Fulbright scholar cut no ice with him, and he forbade me from playing. His reason was that his facility was small, and he wanted no civilian to use his space in competition with the military. Thus, the handball court was not used, and

Captain Brunetti had no one to play with. All other NATO recreational facilities were open to me.

Hiking

During my early handball days at the YMCA, I met many interesting people. Among these was Charles A. (Charlie) Thornton, a long-time Arizonan. Charlie was also a handball player, but not at the level of us younger, more active players. Charlie was about fifteen or twenty years older than I, and he played with a contemporary of his, Andy Tremaine. They were both old-time wrestlers in the southern Arizona scene. Charlie was very bright and interested in many things despite not being educated beyond high school. He was employed as a representative for a local liquor distributing company when I first met him. I knew from conversations with Charlie at the Y that he was a long-time hiker. He was a member of the Tucson hiking club and a good friend of the club's founders, although I do not believe that he hiked with them very much. I, too, was a member of the club, although I never participated in their hikes. I did like their newsletter because it told about their hikes, where they were, how long they were, and so on. When I returned from our Paris sabbatical, I found that my membership had lapsed, and I never renewed it. It was about this time that I started hiking with Charlie on a regular basis. For a long time, this meant about twice a week. On Thursday mornings, we did short local hikes that allowed us to be home by noon. Sunday hikes were longer and more extensive.

Early Thursday morning hikes were often in the Catalinas. This was even before North Alvernon Way had been developed to the base of the National Forest and before the improvement of the Finger Rock trailhead. At this point, we headed up what has now become one of the Pontatoc trails. An old name of the canyon we hiked up was Alamo Canyon. (We called it Alamo #1.) The trail was a remnant of one that serviced the old Spanish mines on the west face of Pontatoc Ridge above us. What was the trailhead for us was at the remains of an encampment that "Old Man Knagge" had built. Knagge ran cattle along the south edge of the Catalinas in the old days, and at this spot he maintained a kennel among other things. He had planted a eucalyptus tree there, and it thrived over the years because of the moisture in the canyon. In fact, in the late 1960s and early 1970s, there was a permanent spring nearby and a small overgrown pool that contained leopard frogs, *R. yavapaiensis*, now extinct from this face of the Catalinas. The spring was being encroached upon by housing developments right up to the National Forest boundary, so I do not know if it still exists. Charlie and I worked on the trail and added small rock cairns ("ducks"). We encouraged friends and acquaintances to use the trail. Often we ascended to the ridge top that afforded a fine view of the *ventana* (window) at the top of Ventana Canyon to the east. Looking back from this point afforded a spectacular view of Tucson and the valley below. I do not remember exactly when it was, but at some point we decided that we needed to devise a return by a different route. We knew that the Finger Rock Canyon trail to the west was not far away, so why not connect the two trails? Linda Vista, a point on the Finger Rock trail, seemed like a logical place for the connection, so we plotted a path through the shin daggers (*Agave schottii*) aiming for Linda Vista. In the course of time, we established a pretty good trail linking Alamo #1 with Finger Rock Canyon. I have no idea if the trail has survived over these last thirty or forty years, but for a time we had a good circular route. It allowed one to park in the lot at the top of Alvernon and to proceed up one canyon and return by another to the same starting point.

After a while, I decided it would be best if I spent less time hiking, so I abandoned hiking with Charlie on Sunday. I should point out, however, that every now and then Lou and I did join Charlie for a Sunday hike. One of the more humorous hikes was to the Santa Rita Mountains. I suggested that we walk to Josephine Saddle between the two peaks of the range. When we arrived at the saddle, and we seemed pretty fresh, I suggested that we proceed a little farther on the Mt. Baldy trail. Lou was agreeable, and we continued on up with the encouraging statement from Charlie of "only five minutes more." Before long, we were at the summit of 9,400-some feet. It was quite memorable for Lou, not only for achieving the summit, but for having done it at the height of the ladybug mating season. This is a time when ladybug beetles by the thousands convene on the summit of nearby mountains to engage in a mating orgy. The problem of being at one of these sites when the little buggers (no pun intended) congregate is that they bite. One can imagine readily how quickly we descended from the summit without even savoring the joy of the achievement.

The encouragement that Charlie provided Lou as we made the Baldy ascent was typical Charlie. He was an accomplished and renowned hiker who did so with ease and without intimidation. He loved to take people hiking who were not experienced or who themselves would never imagine what they were capable of. For years, Charlie organized an annual Palm Sunday hike. He even invited elderly people who were not strong hikers to join the group. This hike was a good five to seven miles in length around the Weaver's Needle in Peralta Canyon of the Superstition Mountains, just east of Phoenix. All who participated really enjoyed the occasion, including Lou and me. There was no pressure, and none of the weaker hikers ever felt that they were holding up the group. It was just Charlie's way—it was a gift he had that was manifested even more strongly in his younger years before I met him. At that time, Charlie coached wrestling at the Arizona School for the Deaf and the Blind. I believe that this was a volunteer activity on his part, but even if it were not, what was a volunteer activity was the hiking he did with these boys. I was really impressed to learn that he took groups of blind kids to the summit of Mt. Baldy. Even more impressive was his having taken these blind boys down to the bottom of the Grand Canyon and back. I met some of these participants in later years and heard about the exploits in their own words.

Charlie and I made the Weaver's Needle hike on other occasions with more seasoned hikers. One of the reasons for the hike was to visit a hermit, Al Morrow, who had a quite comfortable camp in Peralta Canyon. It was located in a glade of tree-like bushes, sugar sumac (*Rhus ovata*), that afforded good protection from the elements. The first hike that we took to meet Al Morrow was made with a friend of ours from the YMCA. He was a young lawyer who was also Al Morrow, but his family was Russian Jews whose original family name was something far different. In any case, when we arrived at the camp, Charlie addressed the hermit with, "Al, I'd like you to meet my friend Al Morrow." The true Al Morrow was so pleased. His family was a New England family of many generations, and of course he wanted to know from which branch of the family our Al Morrow came. Our guy could not bear to tell Al Morrow the hermit the real origins of his name, so he made up some story or other. On later hikes, with or without Charlie, we again met Al Morrow at his camp. When we returned from Italy in 1985, I was sorry to learn that Al Morrow was killed during the heavy winter storms of that year. He had sought refuge in a nearby mine tunnel and the ceiling collapsed on him.

We were in Italy during the summer of 1991, and during that time I was worried about Charlie. He had become depressed as he had aged, and, little by little, his friends had either died or had gotten so deaf that he could not have a decent conversation with them. Moreover, he realized that, at the age of eighty, maintaining a home and yard was not easy. I called him when we returned and found that he had sold his home of many years and had just moved to a retirement community. This had not been a good decision because the denizens of this home were really in bad shape. I called in the middle of the week, and we made plans to meet for a hike the following Monday. I could not contact him on Monday, and later in the day his son Bill called and asked if I had seen or heard from his dad. Of course I had not, so Bill drove to some of the nearby trailheads to look for Charlie's car. He found it at the Linda Vista trailhead on Oracle Road on the next day, and called in Search and Rescue. They found Charlie's body at the base of a treacherous dry waterfall, just up the trail. Obviously, he had fallen to his death. How it happened, we'll never know. What is certain is that many of us lost a good friend.

Subsequently, there was a memorial at which we celebrated Charlie's life. It was attended by many friends and former students who had known Charlie at the Arizona School for the Deaf and the Blind. It was generally a joyous occasion with an exchange of many Charlie stories. Bill Thornton told one that I knew from Charlie's having told me about it at the time. After I had stopped Sunday hikes with Charlie, his son Bill (or Willie as we called him) took my place. Charlie had never met a stranger. Whenever we met someone on the trail, he would always engage the other hiker in conversation. Thus, on one Sunday when he and Willie had hiked Mt. Baldy and were descending, coming up the trail were two young coeds from the university absolutely stark naked except for their hiking boots. As usual, Charlie started talking with them while poor Willie was looking for a tree to hide behind. As I think back over the years, there are so many similar events to tell about. Lou and I have just had the opportunity for a more permanent memorial. In response to a new program, we purchased a net leaf hackberry tree to be planted in Catalina State Park in Charlie's memory. Bill Thornton and his wife, Sue, and granddaughter, Amy, joined Lou and me a few weeks later for the planting.

Being a naturalist at home

Lou and I have had the good fortune of having two homes that fit our needs, interests, and personalities remarkably well. We have lived in our principal home since it was built in 1961 in an old subdivision of the Catalina Foothills. We love it, of course, because it is our home, but it offers an additional dimension of pleasure because it is situated in fairly undisturbed Sonoran desert. For someone with my interests and propensity to be outdoors, our home and its surroundings have been hard to beat. By paying attention, as I have done, there has been much to see over these fifty-plus years. The fact that we are situated on a bend of a large wash that has no easy access other than on foot has provided a nature-watching area many, many times larger than the property we own. There are many species of plants and animals that live on and use this land, but the additional fascination comes from the changes that have taken place over these many years.

We moved into our new home at the end of September 1961. The hot weather was beginning to moderate, allowing me to spend more time outdoors and thus providing a greater exposure to the plants and animals around us. The many species of birds drew much of my attention, especially the year-round residents, since the summer residents had already gone or were leaving, and the winter visitors had not arrived. Aside from their recognition, I began to learn much about our prominent species including, to mention a few, cactus wrens, curve-billed thrashers, brown towhees (now called canyon towhees), verdins, cardinals, gila woodpeckers, gilded flickers, ladder-backed woodpeckers, house finches, gnatcatchers, desert sparrows (now called black-throated sparrows), roadrunners, Inca doves, and black-chinned hummingbirds. Over the next fifty years, there were changes to this list.

One of the early deletions was the loss of desert sparrows from our area, but they are still to be found higher in the foothills. About fifteen or twenty years ago, brown towhees were lost from our environs, but we were recompensated when Abert's towhee joined our family of permanent residents. The differences between these two towhee species are quite striking. While brown towhees are unafraid and almost brazen in their behavior, Abert's towhees are quite shy and are never seen out in the open away from underbrush for long. One can always tell when they are around because they emit a characteristic ringing single note, probably to let their mate know where they are. Couples mate for life and, to be anthropomorphic, seem quite devoted. One of our sad losses was the gilded flicker, a bird that we both enjoyed. As is typical of flickers, they are adorned with a dark triangular patch on their chest, much like a badge. Lou always referred to them as the "Sheriff of Cochise."

I have always had a special affection for verdins, one of our very small species. In part this may be due to a funny incident that occurred years back when Susan Dalby had just moved to Tucson to finish her undergraduate work. She was staying in a dorm, but came up to see us when she wished family-like company. Often that meant she also felt the need to have a taste of nature. One afternoon, she and I took a little walk around our house, and I took her to see a verdin's nest in a tree near the driveway. I explained that the nest is completely enclosed and that the entrance, located on the underside, is protected by what looks like a porch roof. The nest was only six feet or so from the ground and as she stood under it looking at the opening, a verdin flew out practically hitting her in the face. We had a great laugh.

One of our diminishing residents is the roadrunner. When we first moved to the foothills, roadrunners were very common, and we were blessed to have one or two roost on, or very near, the house. Sometimes roosting was a source of amusement. I recall that just a few feet from the window of a room I used as a den, there grew a substantial prickly pear cactus. Late one afternoon, when I lifted my eyes from the desk, I spotted a long black feather extending vertically from the cactus. I concluded that it was a molted feather that the wind had blown in and had become lodged in the cactus. The next afternoon, I noticed that the feather was gone, and I assumed that it must have fallen loose. A few days later, again late in the afternoon, I saw the feather in the same place that it had been before. It finally dawned on me that the feather was a roadrunner tail feather that was in fact still attached to the roadrunner. Over a several-day period, I observed a continuation of this sequence of appearing and disappearing tail feather and realized that the cactus was where a roadrunner roosted in safety from predation. Eventually, by observing the cactus often enough early in the morning and late in the afternoon, I saw the roadrunner wake up in the morning and go to sleep at night. In those early years when we had a lot of roadrunners about, we had few lizards in the patio and around the house. At present, when roadrunners are few and far between, lizards abound.

Two other avian species that have departed from our area are Inca doves and black-chinned hummingbirds. In the old days, black-chins were our primary hummingbird species, sometimes complemented by broad-billed hummers. Now it is rare to see either a broad-bill or a black-chin, although I know that the former does well higher in the foothills and that the latter is abundant in other areas of Tucson. Fortunately, we have two other hummingbird species to grace our area year round. Anna's hummingbird is especially prominent in the winter, and its presence is complemented by Costa's hummingbird that is around most of the year and is especially visible in the spring.

October and November bring the return of winter residents. Among the first to appear are the phainopepla (silky flycatcher) and the mockingbird. While they are both welcome, to have a resident mockingbird is really fun because of the amusing antics it performs and its cheery vocalizations. Actually, the interactions between the males of the two species are also fascinating. Phainopepla males are a glossy black and possess white wing patches that evoke combative responses from mockingbirds that are also marked by such white patches. While phainopeplas do flycatch, especially in the warmer part of the year, they primarily eat mistletoe berries during their tenure in our desert. The result of this diet is a major cause for the spread of mistletoe, since their droppings on branches of shrubs and trees contain both mistletoe seeds and a little fertilizer. It is fascinating to observe a phainopepla perched atop a shrub, and by constantly defecating on this one spot, they build a conical structure of mistletoe waste that grows larger as the season progresses.

One of the long-time winter resident species is the white-crowned sparrow. For most of the years that we have lived in our home, white-crowns were present in sizeable flocks, but as the years have passed the flocks have diminished to only a few individuals. I expect that this change reflects a general reduction in avian numbers. Indeed, in the early years, there were certain species whose winter visits we could always count on and that we looked forward to. These included the two other towhee species of the west, the spotted towhee and the green-tailed towhee. We have not seen a spotted towhee in our backyard in several years and, while the green-tailed towhee still visits every year, fewer and fewer individuals are seen.

The fluctuation in species numbers is not all negative. A prime example is the case of the lesser green-backed goldfinch which, in the early days, we saw only in small numbers in the winter feeding on rosemary plants in the company of house finches. A few years back, I began to see goldfinches in the yard in increasing numbers and in all seasons of the year. They now nest in our area in the spring or early summer. We are certainly pleased to have them, for they are beautiful little birds, and the males call beautifully in the springtime. This is a fine time of the year to take notice of our avian friends as they give voice to the urgency of reproduction. Cactus wrens begin to sing, the assembly call of male Gambel's quail is now accompanied by their "caow call" as they announce their intentions, and the beautiful song of the cardinal, silent in the winter, is now heard. This song is a bit tricky for the call of the male cardinal is quite similar to that of the pyrrhuloxia. Despite the dwindling numbers of cardinals, we have been fortunate enough to have them nest in our yard practically every year. I try to keep an eye open on what goes on in the backyard, and by doing so I am mostly successful in keeping the bronzed cowbird from parasitizing the cardinal's nest with eggs of its own. When I see a male cowbird doing his nuptial performance for his harem of females stashed in the trees, I go out and chase them away. I can't imagine what is so sexy about the male cavorting out in the open pretending to be an avian helicopter.

Over the years, I have looked forward to what I consider to be the true harbinger of spring. It is to hear the first songs of Lucy's warbler together with the sharp calls of the crested flycatchers, the ash-throated flycatcher, and its larger cousin the brown-crested flycatcher. All three species nest near our house and will be around for the summer. Soon after they arrive, I start looking for the beautiful migrants passing through. I focus on our mesquite trees as they begin to leaf out and thus provide food for larval insects. They, in turn, offer meals for such warblers as Audubon's, black-throated gray, Townsend's, hermit, orange-crowned, and Wilson's. These together with the ever-faithful warbling vireo are always a treat to see.

Over the years, we have had the opportunity to see very unusual species in our own yard. It must be about forty years ago when, one winter morning as I left home for the university, I spotted an unusual bird in the pyracantha bush in front of our house. I realized it was a rufous-backed robin from Mexico. I called Lou to see it and asked that she keep an eye out for it during the day because I was sure that some of the bird-watching community would want to see it. That turned out to be the case. It is interesting that, as Tucson has grown, the discovery of out-of-place avian species is practically an everyday occurrence as one can learn from the *Vermillion Flycatcher*, the magazine of our local Audubon Society. There are just so many more bird-watchers out there.

Another unusual sighting took place one Labor Day weekend when I was busy working in our backyard. Suddenly there was a terrible hullabaloo stemming from a telephone pole near our back gate. It seemed as though every small bird in the neighborhood was vocalizing its discontent toward the crossbar of the pole. I assumed that there must have been some sort of a predator there. Indeed, when I retrieved my binoculars to take a look, there was a small prominently tailed owl sitting there and weathering the onslaught. I had never seen a ferruginous pygmy owl, but there it was. When I consulted with ornithological friends, I learned that this little bird-eating owl was very rare. It soon disappeared, and I didn't think more about it. Some time later, as I was working in the yard, I kept hearing a call a little distance away sounding like "poop-poop-poop." It was a call that I did not recognize. Since I pretty much knew all the vocalizations of our local birds, I knew that this was something different. The call persisted for days on end so finally, one Sunday morning, I began walking in the direction of the sound. It took me about a half mile away and up a driveway to the residence of Robert L. Hull, Dean of the College of Fine Arts. We knew Bob and Jeanne Hull, having met them at a reception for University of Rochester alumni. (Bob had at least one degree from the Eastman School of Music.) At any rate, when I met the Hulls at their home and told them what I was after, they remarked that the call was driving them crazy. I told them that I suspected the source of the call was a ferruginous pygmy owl, and indeed we found the culprit.

This interaction with Bob and Jeanne was the beginning of a deep friendship that was partly due to this owl that became quite famous, if not infamous, in the Tucson area. As spring approached, Bob and Jeanne kept us posted on the bird's activities. We learned that the bird had found a mate and seemed to be nesting in a saguaro just below their house, perhaps on the property of Tom Hall, their neighbor and a much-beloved law professor. Before long, the owls produced young and the whole family often perched on a telephone line between the two houses. A secret like this could not be long kept from the bird-watching world, and soon bird watchers from all over the country were everywhere, much to the consternation of the Hulls and the Halls. I believe that this family of owls left the neighborhood after this one reproductive event, but in ensuing years, more ferruginous pygmy owls were found in the desert west of town. Before long this little owl was put on the endangered species list, and, since pairs of owls were found on some prime real estate land, the ferruginous pygmy owl became a *rara avis non grata*!

I have dwelt too long on local birds that have interested me, and as a result have not given other vertebrates, namely reptiles and mammals, the attention that they deserve. Perhaps, at another time or in another life, I can regale a readership about these animals from my vantage point as an amateur.

At the risk of damping any enthusiasm that I may have conveyed about the joys of nature-watching at our foothills home, I mount a soapbox for a few sentences. These words were provoked by the carcass of a gila monster that we spotted on a paved road near our home earlier today. This beautiful reptile died because some careless and thoughtless person drove over it. Perhaps this act was an accident, but I very much doubt it. The gila monster is a diurnal species and, with its size, beautiful orange and black pattern, and slow rate of movement, a driver could not help but see it in the road and thus avoid running over it. Given the number of foothills residents who are absolutely ignorant about the animals around them, my guess is that the driver intentionally committed this act. There are many people who live in the foothills because it is a matter of prestige, and they couldn't care less about the fact that they are living in a unique environment that requires everyone to be a good steward of the land. Unfortunately, today's incident with the gila monster was not a first-time happening. Other reptiles are also victims, but many of these deaths may have been unavoidable. There are not too many of us who will wait for a rattlesnake to cross the road. Or, as Lou and I have often done, get out of the car to nudge a bull snake off the road. On the contrary, we recall that some years back a neighbor of ours, a physician who should have known better, went out of his way to kill an enormous bull snake of near-record dimensions. Lou was following said neighbor up the road when he spotted the bull snake, then drove into his drive where he got out of his car long enough to find a large rock to bash the snake's head. It is a good thing that I was not present! In retrospect, just think of how many pestiferous rodents this snake had consumed during its lifetime and the number that would have met the same fate had the snake not been spotted by our neighbor.

On to a more pleasant theme, for I wish to share some of my naturalistic passions that have been stirred by experiences at our cabin in Greer in the White Mountains of eastern Arizona. In early August of 1973, we purchased a charming little log cabin in the town of Greer. The cabin came with an excellent well and was situated on three-plus acres in somewhat dense woods only a mile from the Greer post office. Greer is at an elevation of 8500 feet, and our place is a bit higher. The forest is composed primarily of ponderosa pine and aspen with a sprinkling of white pine, Douglas fir, and Englemann's spruce. This forest abounds in an array of birds and mammals, together with many plants and beautiful wildflowers. An additional bonus is the presence of edible mushrooms during July, August, and early September. It was all pretty much a paradise for a nature-oriented couple. A drawback was the long distance of 235 miles between our primary home and life in Tucson and the cabin in Greer, albeit the drive is a beautiful one marked by the passage through the spectacular Salt River Canyon. An additional impediment was that we had jobs in Tucson that precluded our spending much time in Greer during the early years. Lou's schedule was always more rigid since she had more or less fixed working hours. Except for scheduled classes, I had more freedom; in fact, when I was involved in a writing project, it was advantageous to be at the cabin. This meant that I was often at the cabin alone, and it didn't take me long to become familiar with our natural surroundings and the denizens of the forest and the valley in which Greer is situated.

Aside from the birds that were always an attraction, I was always on the lookout for whatever amphibians were around. In the beginning, it was the then common North American leopard frog, *R. pipiens*, that was of interest. It was present in such numbers that, on a rainy summer evening, great care had to be taken to avoid running over them as they crossed the road. In those early days, we often walked, or I jogged, to a pond about a mile and a half from the cabin. This is Badger Pond, also near the center of the Greer community. It is fed by springs and is situated alongside the West Baldy trail that leads to the summit of Mt. Baldy, at 11,590 feet, the highest peak in the White Mountains. Badger Pond was a haven for leopard frogs and a principal breeding site in the early spring. Often, I collected eggs, embryos, and tadpoles to take back to Tucson for use in my lab or for teaching. The pond was also a breeding site for a large tiger salamander species, *A. tigrinum nebulosum*. A smaller pond that we always called Crosby's Pond is even closer to town, probably only a quarter of a mile from the post office. In those early years, it was also a breeding site for the leopard frog and the tiger salamander. In fact, it is the location of the last frog egg mass I ever saw in Greer. That was probably in the spring of 1977. I made good use of the tiger salamander eggs that I collected in that pond both for my experimental embryology course and for a research project that Sally (Mason) and I did. It resulted in a fine little paper that Sally wrote.

Aside from walking to various ponds, we, or I alone, did some exploring on forest roads and often I would check out ponds and streams that we found *en route*. One such place was at Three Forks on FR249, a few miles east of Big Lake. Here a large open area is formed where the East and North Forks of the Black River merge with Coyote Creek. In this open area there is a pond fed by an ample warm spring. In the old days, when I got out of the vehicle and trudged along toward the pond, some leopard frogs jumped ahead of me in every direction. They turned out to be young of the year froglets, transformed only a month or two earlier. Moreover, they were not northern leopard frogs such as those at Greer. Instead, they were froglets of what was then called "the southern form," the official designation, *R. chiricahuensis*, had not yet been "fabricated." I knew the frog quite well from my peregrinations in southeastern Arizona. In the pond, I found adults and newly laid egg masses. This was a significant find, as this population became exceedingly important during the next few years. After about two years, John Frost became my graduate student, and these frogs became especially important in his subsequent work on the interrelationships of leopard frog species. In fact, he became co-author of the paper that officially described *R. chiricahuensis*.

After John moved on, my later student, Phil Fernandez, used the Three Forks population for his dissertation research, as I pointed out earlier. From about the time that John Frost joined my lab to the period that Phil was with me, the world-wide loss of amphibians was just being recognized. With respect to Greer, there was a cataclysmic loss of leopard frogs in about 1978 that led to the disappearance of both *R. pipiens* and *R. chiricahuensis* from most localities in the White Mountains. An exception was the Three Forks population. Thus, Phil was able to procure the small numbers of frogs that he needed for his dissertation research. During the first few years after Phil had finished his PhD and taken a position at Grand Canyon University, he would often drive up from Phoenix, and we would try to discover extant populations of frogs. We found no frogs, but we did observe that all the classic sites had, in common, the presence of considerable numbers of crayfish. I had observed crayfish in certain ponds in the Greer area in 1973 and had assumed that they were native species. When Phil returned to Phoenix, he did appropriate literature research and learned that crayfish were not native to the state. As Phil and I explored former frog sites, what stood out to me was the total absence of vegetation in the water. Instead, these ponds and streams had a gray, sterile appearance. With no vegetation, tadpoles could not grow; thus, while the introduction of crayfish may not have been the cause of the demise of frogs, their populations could never be restored because of the absence of food.

The crayfish story is one that I should not go into here, but it should be pointed out that the Three Forks area was eventually invaded by crayfish and that all frogs were soon gone. Fortunately, Phil had established a breeding population of frogs from Three Forks in a greenhouse at Grand Canyon University. If I didn't teach him anything else during his tenure in my lab, he at least learned how to manage a breeding colony successfully. Phil had established fine relationships with Arizona Fish and Game people. These folks have followed his advice and used some of his frogs to help establish a breeding colony of Three Forks frogs at the Pinetop office, to be used for the eventual reintroduction of this frog in the White Mountains. Before finishing this frog/crayfish story, I should point out that, with grant support, Phil was able to prepare a report of the crayfish problem. This report received the attention of many, including our local PBS station, KUAT. They wished to film a segment on crayfish for their program, *The Desert Speaks*. Phil referred the station to me in the late summer of 2003, when I was still at Greer. I consented to show them where the crayfish had invaded Three

Forks, and thus I took the filming crew out to film a segment, "Unwelcome Guests," that was aired on KUAT in 2004.

A propos of my having often gone to the cabin alone during our pre-retirement years, a rather fascinating wildlife event took place one evening back when I was there by myself. During those brief stays, I ate as simply as possible and took advantage of the fact that I have maintained a sourdough starter for many years. One day, I put up a sponge (*biga*) in preparation for baking a focaccia for my supper. My version of this flat Italian bread includes a topping of a mixture of two cheeses, fontina and mozzarella, together with herbs and kalamata olives that are put in place just before baking. The aroma of the baking focaccia permeated the cabin and was still present after supper when I sat down to read. The quietude of the cabin was interrupted by a noise at the back door, our primary entrance. At that time, we had not yet installed an inside switch for the back door light, and one had to go outside to turn the light on and off by a drawstring. As I approached the back door, I had the good sense, fortunately, to pound on the door before I opened it. When I did, there I was, face to face with a black bear! In a reflexive move, I slammed the door. I had no idea how big this bear was, but I knew that it was a lot larger than I!

The next morning when I opened the curtain on the window next to the door, I found the calling card the bear had left. In the center of the pane was an imprint of the bear's wet nose complete with a pair of nostrils from which phlegm had dribbled down. Toward both edges of the pane, complete paw marks had imprinted. Of course, I photographed this evidence for posterity, but as additional assurance, we did not wash that window for many months. I have had other encounters with bears, but never as intimate as this one.

I cannot leave my recounting of natural history at our Greer home without reference to the avian world. However, there is so much to tell about our experiences with the birds around our cabin, that I'll say nothing other than that I have viewed them with the same passion that I expressed about our Tucson birds.

Epilogue

I have always felt that an epilogue was partly meant to dampen the inertia developed in writing the preceding chapters. As I arrive at this point in my writing, that sentiment is reinforced, although there are some additional points to be made. One of these relates to mortality, for since I started writing this memoir, one of my dear friends, a magnificent man, Howard Bern, passed away at the age of ninety-two. Howard was an extraordinary person who gave to so many people. He helped me in many, many ways. Just a month or so ago, my former student, John Frost, succumbed to ill health, probably a heart-related event. John was too young to depart this earth, and he did so without fully realizing his potential as a scientist and teacher. If only he had better taken care of himself. John was the second of my graduate students to precede me to the grave. The first was Lee Stackhouse, whose death was tragic and unexpected. He and his wife, Bernice, perished in the crash of a commuter airline *en route* from Indianapolis to Chicago in 1994, at the height of their business success. How strange it is that he and John had a unique relationship among my former students, for, if it were not for Lee and Southwest Scientific Supply, I would never have met John.

I should mention my own mortality. I am here and do not expect to go anywhere in the near future. When I do succumb, it can be looked at as unfinished business that is at last completed. My life will be completed then and there, and all that will be left will be the physical elements that had composed my physical being. I will have left my writings and memories that others may have. More important, I will leave behind the love that has bound Lou and me for all these years. Should she survive me, she will not easily forget her mate and the life we had together.

But, what about my soul, and the hereafter, and all those things that concern many people? They are of no concern to me, because, for me, they do not exist. We are merely elements in a continuing evolution in a universe that continually expands by the second. Perhaps in future millennia, our descendants will be able to explain how it all came about. I am confident only that there was a cause, because I believe in causality. However, I am not arrogant enough to personify that cause. As for the hereafter, I believe that the hereafter can be best explained by that vintage joke about the cleric visiting an aged parishioner. It seems that when the parson arrived at the home of Mrs. Jones, he asked her to think about the hereafter. She responded that she often thought about the hereafter. In fact, invariably when she went to the pantry, she always asked herself, "Now what am I here after?"

Acknowledgments

One of the pleasures that the writing of this memoir has given me is another opportunity to interact with friends who have helped as the manuscript developed. Lou, who has been a critical copy editor of my writings over our long married life, has again helped. Susan Eastman, who provided secretarial support before each of us retired from the Department of Anatomy, kindly played this role as the manuscript moved to completion. She has been an invaluable help as a cat sitter and house sitter over the years and we cherish her friendship. Many others have helped in various ways from the beginning, and I acknowledge their contribution. They include Jenny Felber, Linda Stitt, Vincent Hearing, Paul Maseman, Lucinda Davis, John Gentile, Robert Erickson, Rakesh Rastogi, Luisa Iela, Jiro Matsumoto, Ray Moldow, and Susan Dalby.

A Selected Glossary

Agouti Signaling Protein a peptide that influences mammalian pigmentary patterns by blocking the interaction of MSH with melanocytes, thus inhibiting eumelanin biosynthesis.

Animal Pole the upper half of eggs that are polar due to the unequal distribution of yolk. This pole, often heavily pigmented, contains less dense cytoplasm than its opposite, yolk-laden Vegetal Pole.

Buccal Endoderm the portion of the embryonic mouth derived from endoderm, the innermost germ layer.

Chromatography a chemical procedure that allows for the separation of constituents of a mixture.

Chromatophores pigment cells of lower vertebrates and invertebrates, often involved in color change.

Corticosteroid a steroid hormone produced by the adrenal cortex.

Carotenoids a group of yellow to red pigments of plant origin composed of chains of isoprene units. Two subdivisions are **carotenes** and **xanthophylls.**

Cyanophore	a chromatophore containing a blue pigment, so far found only in a few species of fish.
Dermal Chromatophore Unit	a contiguous group of dermal chromatophores (melanophore, iridophore, and xanthophore) that function together in the color change of many amphibians and reptiles.
Determination	a term from classical experimental embryology used to designate when the developmental fate of an embryonic part or region is sealed, viz. eye, limb, pigmentation pattern.
Ectoderm	the outer of the three embryonic germ layers from which an embryo is derived. It gives rise to the skin and nervous system, while **mesoderm**, the middle layer, gives rise to muscle, the skeletal system, and parts of internal organs. The **endoderm** is the origin of the gut and enteric organs.
Embryonic Induction	the phenomenon whereby one part of an embryo induces a contiguous part to differentiate into a specific tissue or organ. The ability to respond to the inductive stimulus is known as **competence.**
Expansion— Contraction	same as dispersion—aggregation. The capacity of a chromatophore to respond to stimuli by causing their respective pigment-containing organelles to migrate to the periphery or to the center of the pigment cell.
Flagellum	a whip-like structure possessed by some cells causing them to be motile, as in a spermatozoan. Sometimes the sperm cell flagellum may have an attached **undulating membrane.**

Fertilization Membrane	also called a **vitelline membrane**, lifts off the egg cell surface at fertilization. The space thus created between the vitelline membrane and the egg surface is the perivitelline space.
Gonadotrophin	one of a series of pituitary hormones that stimulate the ovary or testis, viz. **Follicle Cell Stimulating Hormone** (FSH), **Luteinizing Hormone** (LH), **and Prolactin.**
Guanophore	an historic name for **iridophore**.
Hypophysis	the pituitary gland, especially in embryonic stages. Its extirpation is referred to as **hypophysectomy**.
Intermedin	an historic name for **Melanocyte Stimulating Hormone** (MSH), so called because it derives from the intermediate lobe of the pituitary gland.
Iridophore	a reflecting pigment cell or chromatophore that reflects light and appears iridescent.
Melanophore (Melanocyte)	a lower vertebrate chromatophore that contains black or brown melanin pigments, notably eumelanin. In birds and mammals it is called a melanocyte. Some melanocytes may be red due to their capacity to produce phaeomelanin, a reddish melanin pigment.
Melanogenesis	the biosynthesis of eumelanin and/or phaeomelanin.
Melanosome	the intracellular organelle that contains the melanin pigments.
Melatonin	a simple indole hormone of the pineal gland.

Mosaic Chromatophore	mixed pigment cell containing two or more unrelated pigments, sometimes within the same organelle.
Neural Plate	a flat, precisely defined, elongated plate of ectoderm on the dorsal part of an embryo that is delineated by neural folds. During development, the plate rolls up to form the neural tube that forms the central nervous system. The front end becomes the forebrain followed by the midbrain, hindbrain, and spinal cord.
Neural Crest	a string of embryonic cells of neural fold origin that sit atop the closing neural tube. Neural crest cells migrate remarkably throughout the developing embryo to give rise to an array of derivatives from ganglia to pigment cells, to teeth, to endocrine cells, and so on.
Optic Vesicle	one of two, left and right, bulbs that grow laterally from the embryonic forebrain to contact the outer ectoderm there inducing lens formation. Optic bulbs give rise to the retina and its associated layers, and so forth.
Pineal (body and gland)	an organ derived from an embryonic outpocketing, the epiphysis that pushes out from the dorsal midline of the forebrain. It produces melatonin in accordance with a daily rhythm.
Pteridines (pterins)	double ring compounds often used as pigments in lower vertebrates and insects. Their prevalence in butterfly wings gives them their name. They are related to **purines**, other double ring compounds such as guanine.

Stomadeum	the early mouth of an embryo that includes the oral plate (or membrane) that covers the entry to the primitive gut.
Visual Pigments	photolabile pigments related to vitamin A that are located in the rods and cones of the retina and that are the basis of vision. Rhodopsin and porphyropsin are noted visual pigments.
Vitellogenesis	a process of the liver that produces vitellogenin, a large protein that is deposited in eggs of species as varied as butterflies and birds. Vitellogenin is transported from the liver to the ovary in the blood serum.
Vitiligo	an abnormality of pigmentation in some vertebrates, notably man, characterized by the presence of white patches of skin. Lesions of vitiligo are characterized by the localized loss of melanocytes.
Xanthophores	yellow pigment cells, especially of fishes, amphibians, and reptiles. Pteridines and/or carotenoids are their typical pigments. A red pigment cell equivalent is the erythrophore.

Especially Relevant Research Publications

Bagnara, J.T. 1957. Hypophysectomy and the tail darkening reaction in *Xenopus. Proc. Soc. Expt. Biol. Med.*, *94*:572-575.

Bagnara, J.T. 1958. Hypophyseal control of guanophores in anuran larvae. J. Exptl. Zool., *137*:264-284.

Bagnara, J.T. 1960. Pineal regulation of the body lightening reaction in amphibian larvae. *Science 132*:1481-1483.

Bagnara, J.T., J.D. Taylor and M.E. Hadley. 1968. The dermal chromatophore unit. *J. Cell Biol., 38*:67-79.

Bagnara, J.T. and M.E. Hadley, 1973 *Chromatophores and Color Change: The Comparative Physiology of Animal Pigmentation*. Prentice Hall, Inc., Englewood Cliffs, New Jersey.

Bagnara, J.T., J.D. Taylor, and G. Prota. 1973. Color changes, unusual melanosomes, and a new pigment from leaf frogs. *Science, 182*:1034-1035.

Bagnara, J.T., W. Ferris, W.A. Turner and J.D. Taylor. 1978. Melanophore differentiation in leaf frogs and the deposition of a new pigment. *Dev. Biol. 65*:149-163.

Bagnara, J.T., J. Matsumoto, W. Ferris, S.K. Frost, W.A. Turner, Jr., T.T. Tchen, and J.D. Taylor. 1979. On the common origin of pigment cells. *Science, 203*:410-415.

Bagnara, J.T. 1982. Development of the spot pattern in the leopard frog. *J. Exptl. Zool.*, 224:283-287.

Marinetti, G.V. and J.T. Bagnara. 1982. Yolk pigments of the Mexican Leaf frog. *Science.* 203:410-415.

Bagnara, J.T., L. Iela, F. Morrisett, and R. Rastogi. 1986. Reproduction in the Mexican leaf frog. *Pachymedusa dacnicolor.* I. Behavioral and Morphological aspects. Occasional papers of the Mus. Nat. Hist. Univ. Kansas, Lawrence Kansas No. 121, pp. 1-31.

Bagnara, J.T. 1987. The neural crest as a source of stem cells (pp 57-87. *IN:* Developmental and Evolutionary Aspects of the Neural Crest, P. Maderson (ed.) J. Wiley Sons Inc.

Fukuzawa, T., P. Samaraweera, F.T. Mangano, J.H. Law and J.T. Bagnara. 1995. Evidence that MIF plays a role in the development of pigmentation patterns in the frog. *Develop Biol.* 167:148-158.

Bagnara, J.T. Chapter 2: Comparative Anatomy and Physiology of Pigment Cells in Nonmammalian Tissues. 1998. In: "The Pigmentary System—Physiology and Pathophysiology." (J.J. Nordlund, R. Boissy, V. Hearing, R. King, J-P. Ortonne, eds.). Oxford University Press (New York). pp. 9-40.

Bagnara, J.T. 2003. Enigmas of pterorhodin, a red melanosomal pigment of tree frogs. Pig. Cell Res. 16:510-516.

Bagnara, J.T., P.J. Fernandez, and R. Fujii. 2007. On the blue coloration of vertebrates. Pig. Cell Res. 20:14-26.